# WILLIE ROBSON
## HIS WORDS

Willie Robson

Northern Bee Books

Willie Robson - His Words
© Willie Robson

All rights reserved. No part of this publication may be reproduced, stored in a retrieval system, transmitted in any form or by any means electronic, mechanical, including photocopying, recording or otherwise without prior consent of the copyright holders.

ISBN 978-1-914934-48-3

Published by Northern Bee Books, 2022
Scout Bottom Farm
Mytholmroyd
Hebden Bridge
HX7 5JS (UK)

Design and artwork
DM Design and Print

# Willie Robson
## His Words

Willie Robson

# Contents

*Foreword* .................................................................................................... 1
*Black bees* ................................................................................................. 2
*Swarm at a Wedding* ............................................................................... 4
*Honeycomb, Pollen and Brood* ................................................................ 6
*Trees* ......................................................................................................... 9
*Honey bee viruses* .................................................................................. 11
*A discussion on swarming assuming the bees have enough space* ...... 12
*June Clover* ............................................................................................. 15
*Clutching at straws* ................................................................................ 17
*Himalayan Balsam 2021* ........................................................................ 18
*Beowulf Cooper* ..................................................................................... 19
*Further concerns 2018* .......................................................................... 21
*Beekeeping notes for September 2019* ................................................. 22
*Robbing* .................................................................................................. 25
*Practical Beekeeping* .............................................................................. 27
*Hives cannot function without a queen* ................................................ 31
*Insecticide in London Parks* .................................................................. 33
*Concerns about the health of our bees* ................................................. 35
*Observations from Willie Robson* ......................................................... 37
*The Smith Hive* ...................................................................................... 43
*Black Bees – Decisions critical to their survival* .................................. 46
*Viruses, Varroa and Dilution* ................................................................. 48
*An Assessment of the Black Bee going back 100 years* ....................... 50
*Andrew Scobie obituary* ........................................................................ 54
*Selby Robson* ......................................................................................... 56
*Willies Memoirs – Early Years* ............................................................... 58
*Willies Memoirs – Our House* ............................................................... 62
*Willies Memoirs – The Bees* .................................................................. 67
*Willies Memoirs – 1985* ......................................................................... 69
*Willies Memoirs – Supermarkets* .......................................................... 70
*Willies Memoirs – Cosmetics* ................................................................ 72
*Willie's Memoirs – The Showroom* ....................................................... 73
*Willie's Memoirs – Tractors* ................................................................... 75
*Willies Memoirs - Conclusions* .............................................................. 76

# *Foreword*

These articles were written in a hurry because I was responsible for running the business as well as looking after the bees. There is a considerable degree of repetition for which I apologise. There is also some useful information which will not appear in textbooks.

At the same time, I thought it was worthwhile making a record of the establishment of our honey farm otherwise I might regret not doing it.

# Black bees

I am surprised to learn that a small sample of our black bees have shown themselves to be 90-94% *Apis mellifera mellifera* as a result of genetic analysis undertaken by Dr Mark Barnett of Beebytes Analytics CIC.

During the 1950's very many Italian queens were imported into this country and so it was very normal to see bees with yellow stripes among colonies of black bees. Since then, there have been imports from very many countries into an area with a cool climate (the Eastern Borders).

Traditionally our black bees superseded at the heather when the old queen was 3 years old. I believe that the indigenous queens and drones will come out in poor weather and get mated right beside the hives because it is the only way that they will survive another year.

Crossbreeds/imports will be reluctant to come out at such low temperatures and thus their genes are eliminated over a long time (we know this because the yellow colours have disappeared from them). Thus, the black bees become inbred but not to the extent of being useless, far from it. I would hazard a guess that the black colour is to absorb sunlight on a cold Spring day when pollen is needed in the hive. Thrift is to get through the bad times without starving and to keep a manageable broodnest. Fifty years ago, there were many feral colonies some in hives but mostly in the trees. They would be 100% *A. m.m.* but very different in characteristic to a domesticated black bee being good pollinators but useless for honey production. Beekeepers, ourselves included, did struggle with them because of swarming and general negativity. Brother Adam would not think much of them, but they would have some useful genes. Scientists in Austria predicted, before varroa arrived, that true ferals would be the only bees to survive varroa unaided. They were right to a degree.

True ferals or 'wild' honeybees are smaller and sometimes hairy (throwback) and keep a small broodnest. The queen cells are numerous and tiny and often built horizontally and hidden. A domesticated colony will build queen cells that are proud and vertical. Feral bees will move across the face of the comb in waves to get away from the light and the queen will hide. A good black bee will sit patiently on the face of the comb unless they have had too much smoke. These are useful indicators.

Progressive beekeepers would select the best of the ferals and work away until they had a good commercial strain of their own. WW Smith of Peebles was perhaps the most notable. Imports were not necessary. However, the grass is often greener, and imports were brought in and the resulting out crosses sometimes became unmanageable (aggression). Beekeepers can build a strong relationship with their own black bees over time with sympathetic management to the extent that should the beekeeper pass away, the bees will deteriorate very soon afterwards. No-one will take them over and be as successful.

I should say that black bees will not tolerate rough handling without becoming habitually defensive. This gets them a bad name. Colonies of black bees within an apiary will all be at different stages and will work as individuals (drifting). They are often suffering from chalk brood. This could be due to malnutrition. They are otherwise tolerant (partially resistant) to most other diseases but not modern viruses.

I have tried to describe all the variations of a black bee that might be found under the general description of an indigenous black bee. Ideally, they need to belong to an individual with ability. Real ferals certainly do not belong to anyone. When defining black bees these distinctions are important.

# Swarm at a Wedding

We went to a wedding knowing that a swarm of bees was hanging high up on the 11th century church. I left the empty hive on a dustbin and, when we came out, they had taken up residence.

The swarm would be hanging on the church because the 'scouts' would have identified a place to start a new home and when they got there it was not apparent where the entrance was. When I put the hive on the dustbin the 'scouts' would be relieved to find it and bring all the 'repletes' in.

A swarm will only leave their parent colony if there is an imbalance within the colony. I.e. too many young bees being born and not enough work for them. The colony of honeybees then becomes 'smitten with the bug' as they enter a predetermined 'state' where they make preparations to swarm.

The young bees, for whom there is insufficient work, fill themselves up with honey and they stay in the hive (colony) so they are unable to orientate on the present site. This is so that when they fly with their queen to their new home a mile or two away so they cannot find their way back to the parent hive. They are known as repletes.

They are guided to their new home, in this case the church roof, by older bees called scouts. These older bees emit a pheromone to which the bees are attracted so they know where they are going. The pheromone consists of geraniol, nerolic and geranic acids and citral and is emitted from the Nasonov gland within the abdomen of the adult bees. The queen emits a substance (another pheromone) called 9-Hydroxy-2-enoic acid which holds them all together as a group (the swarm). Additionally, the scouts, when they have found the entrance to a good home, fan their wings making a loud noise which indicates pleasure and contentment. The swarm is then shepherded to the correct location by a mixture of pheromones and sound, otherwise they would get hopelessly lost. They carry enough food with them in their honey sacs to tide them over for a fortnight as an insurance against bad weather. They are not aggressive or defensive at this time, far from it. This is extremely sophisticated behaviour.

Swarming is the reproduction of the colony in order to cover new ground (expansion) as well as replace colonies that have died out in the winter. A honeybee colony is corporate i.e., they function as a single individual. This is about decision making within the colony. Decisions have to be made every day which must involve intelligence as well as instinct.

The viruses that are affecting honeybees now are causing honeybees to lose their instincts to some degree. Agricultural chemicals may or may not be involved. I cannot say for certain. Consequently, swarming is greatly reduced and colony numbers are greatly reduced.

If indigenous bees were not looked after by beekeepers, they would die out altogether. Pest control officers in North Northumberland tell me that the incidence of honeybee swarms has gone down by 90% in the last 20 years. This is a dire situation and was never the case in the past. My Father said if the bees were swarming, "All was well in the world" and that was fifty years ago. He was right, swarming was a cause for celebration both for the bees and the beekeeper. Not everyone saw it that way.

# Honeycomb, Pollen and Brood

Honeycomb is made of beeswax. Beeswax is made by young bees from the nectar of flowers. Generally they make it during hot weather in the Summer when there is a 'honeyflow' and nectar is being secreted by flowers. The wax is kneaded into honeycomb which is then used to store the honey which is being produced at the same time. Six sided cells are made that interlock, each cell wall is also the wall of the adjoining cell. Similarly, the bases of the cells are interlocked with other cells that face the opposite direction to form a 'comb'. The cells are inclined at 15° up from the horizontal so that surface tension prevents the honey from running out of the cell. The surface tension depends on the viscosity of the honey which in turn depends on the moisture content which has to be below 18%. When the moisture content is below 18% naturally occurring yeasts are unable to live because honey is hygroscopic and draws moisture out of living organisms (including bacteria which of course makes it an ideal dressing for wounds). Thus, the honey in the cells cannot ferment so it will keep for years and provide the bees with food for many Winters.

Honey is very heavy, half as heavy again as water. Beeswax is very light and yet a few ounces of beeswax can support many pounds of honey because of the structure of the honeycomb. Honeycombs are built about 1½ inches apart and are separated by a 'bee space' which is always the same allowing the passage of a bee (less than ⅜inch). This measurement never varies; indeed, all the measurements and geometry are invariably correct showing that the bees have mastered the technology necessary to construct a very strong honey comb out of very little material. They also understand the principle of surface tension and the importance of low moisture content in the honey. They are super sophisticated. They have mastered all this 'technology' thousands of years before man.

Honey comb is also used by the bees to store pollen (protein) and as a nursey for young bees (pupae).

## Honeycomb and brood

I have mentioned that the bees use honeycomb to rear their young (brood). The brood area is always at the centre of the hive where it can be kept warm in sub-zero temperatures. May 2021 saw 21 nights of frost which must have put a great strain on honeybees trying to keep their brood warm.

During the Summer honey bees live about 6 weeks with their duties divided into housework (honey bees collect propolis from the trees as a resin containing bioflavonoids. They polish the honeycomb with it particularly in areas where the young are reared in order to 'sterilize' the environment within the hive and prevent growth of bacteria. The interior of a hive is 'spotless'), tending to the brood and finally foraging for water, pollen and nectar (honey). There might be 50 000 individuals in a hive, and they are all replaced within 6 weeks. Vast amounts of food have to be brought in to build all these bodies and to accumulate enough honey to see them through the Winter.

During the Winter the worker bees need to live for 6 months or maybe 7 depending on the Spring. They change their physiological state to sit out the bad weather without hibernating. A few bees will always be active. The varroa mite which has plagued our bees for 20 years (and worldwide bees for 40 years) has introduced viruses into the colonies which have affected the bee's ability to prepare for the Winter. They then suffer from 'reduced life expectancy' and perish in the late Winter.

Varroa can be controlled to some degree by the use of hard chemicals, but we use organic treatments (oxalic acid and thymol). Beehives already contain too many chemicals associated with intensive agriculture without putting any more in. Keeping bees has become very difficult much more so than when I started 60 years ago.

## Honeycomb and pollen

I have mentioned about bees storing pollen in honeycomb. Pollination is about bees visiting flowers to gather pollen to feed their young principally, as well as themselves. They provide a vital service to plants by transferring pollen from the anther in a flower to the stigma in another flower, often some distance away, thus achieving fertilisation and ultimately seed that will be the start of a new generation of plants. Some plants are self-pollinating and others, like grasses, are wind pollinating. The pollen collected by bees is the protein part of their diet and is carried in a basket (corbicula) on their hind legs. Honey bees will make great efforts to collect pollen long before Spring in order to give the young grubs the very best nutrition. They have been seen to pollinate Christmas roses when no other insects are around. If the temperature is above 6°C and the sun is out they will go. Experts tell us that there are more efficient pollinators than honey bees but they forget that honey bees are reliable pollinators 12 months of the year because they do not hibernate. There are always one or two bees on the outside of the cluster that can detect intruders as well as pollen at a considerable distance away. Thus, they provide a vital service to plants early in the year and throughout the year. There would be no almonds if it were not for honey bees.

This information was written primarily for the general public.

# *Trees*

I was thinking the other day about the sycamore tree and the other trees that will soon be flowering.  There are already tassels laden with pollen on some of the willows and the alder will also provide plenty of pollen.  Much more prominent in the landscape will be the blackthorn (sloe) in March and the gean (wild cherry) in April.  There must be blackthorn bushes grown commercially because shop shelves are laden with sloe gin.  During a warm spell in April, bees will get some surplus from all the different varieties of "prunus" but generally the honey will be used in brood rearing.

In days gone by, we all waited for the sycamore to flower in the hope of a crop.  All too often harsh winds blew the tassels off before the bees could get going.  However, if there was a period of settled weather thereafter, the bees could get a respectable crop of sycamore honey from the "extra floral nectaries".  This was delectable honey of a greenish red colour (engine oil) and always of show quality.  There were woods in the Lothians that contained many mature sycamores that might provide a reliable honey crop.  When oilseed rape appeared in the early 70's that was the end of sycamore honey in the East unfortunately.  Rape honey was easy for the bees to collect.

We are told by those who know better that the sycamore tree is a weed and should not be planted.  I remember that some years ago timber merchants were scouring the countryside for the trunks of sycamores, particularly those that were rippled, the rippling within the growth rings being caused by unrelenting pressure from the prevailing wind.  One thousand pounds was being offered for a suitable tree that would be exported to Germany to make furniture.  There we have the sycamore, a tree that will provide nectar and sustenance for many insects and that in turn will provide food for birds.  We have noticed that when a cold East wind blows in from the coast in mid-summer, many swallows will be flying in the lee of sycamores to feed on insects that have been dislodged, a bonus for the swallows on a bad day.

I realize that I have forgotten about the elm that flowered profusely before the leaves appeared.  Passers-by would remark that there must be a swarm of bees in an elm tree in March because there was so much activity.  We rarely see an elm tree now and soon all the ash will have disappeared.

The next crop to look forward to was the hawthorn (May blossom) that flowered for 10 days only.  All the hedges in Northumberland were 20ft high mature hawthorns, providing shelter for cattle.  Bees might get a super of highly aromatic medium colour honey, one year in ten. These hedges were grubbed out to make way for cereals using government money.

Lime trees are grown in avenues attached to big houses and provide some nectar in most years and occasionally a good amount. Limes flower at a time when bees are consolidating rather than rearing brood. There is a lot of lime honey produced in the West of Scotland and there are many different varieties of limes.

Lastly, we must mention *Thuja plicata* (Red Cedar) that provides ideal timber for beehives. We know about Red Cedar from Canada but not so many know about Thuja grown in this country because it is not so readily available, being grown as an amenity tree rather than in plantations. The British tree provides coarser timber but stronger and always dimensionally stable and good for a fine finish. It is exceptionally resistant to decay provided the sap wood is not used and very light, perfect for beehives. We have used many tons, bought the trees, dried the wood and made hives that will last a generation or more.

Reading through this it is easy to see that 'loss of habitat' is very real. A small oak wood was felled near an apiary a year or two back. Fair enough but I thought about the number of (rare) birds using that wood and all the insects specific to oak woods and multiply that by decades and think about the loss in terms of habitat.

Honeybees will visit oak trees, beech as well, in the spring maybe, but much more likely in August where they will gather "honeydew" which may be seen at honey shows in the very dark classes!!

Paul Robeson sang a song entitled "Trees". The tune will be familiar to many of our senior generation…that includes me!

# *Honey bee viruses*

Honey bee viruses are commonly detected in colonies free of any recognisable disease signs and healthy honey bee colonies undergo constant cycles of viral infection (Highfield et al, 2009). Colony losses are often attributed to the co-occurrence of varroa destructor and viruses (Schroeder and Martin 2012). Up until the 1950's the ectoparasite mite varroa remained a parasite of its natural host, Apis cerana, the Asian honey bee which is able to control mite populations to a tolerable level. However, varroa has since shifted host to the Western honey bee, Apis mellifera. Varroa has been identified as one of the major factors responsible for colony collapse, having been linked to the global collapse of millions of honey bee colonies by changing the honey bee viral landscape (Martin et al 2012). One virus in particular, deformed wing virus (DWV) is a re-emerging epidemic that has been sustained by varroa. Varroa feeds on the haemolymph of honey bees providing an alternative transmission route for DWV which then replicates to high levels, eventually leading to colony loss.

We are aware that massive diversity exists within DWV. In addition honey bees are often also infected by more than one species of virus, acute bee paralysis virus in particular and two others.

Although honey bees are exposed to multiple threats of habitat loss and pesticides it is generally accepted that the main cause of honey bee population decline is infection by their viruses and its association with varroa. Therefore, the degree of varroa coupled with the diversity of viral infection could be the best indicator of honey bee colony health.

This has been copied from an article by Declan Schroeder of the Marine Biological Association and tells us where we are now.

Our best advice going forward is that we must treat our colonies more often with oxalic acid or thymol (organic) to eliminate every varroa mite. This is contrary to good husbandry which would suggest that we try to get our bees back into total natural immunity. There are other chemicals that would control varroa but they are of a type which would be better not used in a beehive. However, these chemicals are keeping beekeeping going worldwide.

WS Robson

September 2021

# A discussion on swarming assuming the bees have enough space

Bees will swarm for several reasons. Firstly, there is absconding which is caused by the sun beating onto the front of the hive when the wind is in the South. No queen cells are built until after the swarm has left (Eggs in queen cups) or emergency cells thereafter.

Then there is habitual swarming, every year, same week, (section bees, true ferals and purebred Carniolans).

Thirdly, there is swarming due to chronic instability when the bees build queen cells all summer and never settle down, being ill at ease with themselves for various reasons. They are not true to type, i.e. perhaps too many outcrosses with imports that are not indigenous and not progressive and stable.

A failing queen will cause a colony to swarm in order to replace her. One or at most two queen's cells denotes supersedure i.e. no swarm.

Fifthly, there is normal swarming, which involves the colony becoming "smitten with the bug" in order to adhere strictly to the steps needed to make a success of the entire procedure until there are young bees hatching in the new site. The parent colony becomes 'smittle' for 2 reasons. Firstly, there will be colonies within a 2-mile radius that have died out in the winter and these need to be repatriated right away to maintain colony numbers. Secondly, in the spring, queens are laying eggs on a daily basis and there comes a time when more bees are hatching than are dying through hard work. This situation is made worse in the UK by variable weather patterns which stops them working for several days or longer so there is imbalance within the colony i.e. too many bees ready for work and too many older bees for whom there is no work due to bad weather or the June gap. The resulting overcrowding causes them to become 'smittle'.

They then push all the young bees up into the supers and corral them there until the swarm is about to leave. These bees do not work, and neither are they allowed to fly and mark their hive. They are full of honey and are known as 'repletes'. They intend to take the honey to the new site. It was possible to guess which hives were going to swarm by the behaviour of the bees in the supers. If the bees were building wax in the supers (we were using starter strips) then it was unlikely that they were going to swarm. If they were chewing at the foundation or reluctant to prepare the combs to receive honey, then they were about to swarm. Every beekeeper will have seen this. Sometimes a box of foundation will make them more determined to swarm (taken the huff). They are only going to make wax in a new home. Some swarm sites will have to be built from scratch before the queen can lay an egg so the repletes do not engage in any work that might tire them or use up any of their reserve. Honeybees are quite exceptionally sophisticated, as we know.

When the swarm leaves the hive, always on a good day, the repletes go together with many old bees and a great deal of excitement and happiness. Nobody gets stung.

Thereafter some of the old bees return within half an hour, as do a few more in the following days, especially if there is heavy honey-flow. The old bees possibly forget what they are supposed to be doing. They need to prepare for a second swarm with the first young queen that emerges. In some cases, the old bees that go with the second swarm lose interest right away and leave the repletes to fend for themselves and they die of starvation. They cannot return to the parent hive because they do not know where it is. In the autumn one can see where they have been hanging in a bush. A few combs bleached white and abandoned. This is a blip in an otherwise very well-rehearsed procedure.

During swarming evolution has built in some options, the management team may have a change of heart. This is generally caused by a rise in atmospheric pressure and the prospect of a heavy honey-flow. They are acutely aware that at 56 degrees North Winter is coming and a crop of honey is needed. Otherwise very bad weather may cause them to change their mind or they may swarm at a later date with a virgin queen. This tells us that there is corporate management within the colony and instant communication, very like an army in the field. Each colony is an individual with its own leadership. We have seen them out on a branch preparing to leave and the old bees lead the repletes back into the hive and resume work and forget about swarming. This would not happen in a colony that was unstable.

Currently we have one colony in a sheltered village where the bees are constantly busy and occupied with the queen providing only enough brood to make up for the losses caused by old age. Thus, the colony remains in perfect balance and no swarming occurs. If a swarm does emerge it goes straight back. Confidence within the colony is further improved if a full super of honey/pollen is left above the brood chamber permanently (brood and a half). Then they might replace their queen by supersedure. Colonies that supersede over many years may lose their vigour through inbreeding.

Please note, all these 'activities' will have been suppressed within modern hybrid colonies but not totally. The above article has been written for general interest only and has little relevance to present commercial beekeeping.

A further discussion about swarming and the artificial swarm.

In order to make an artificial swarm the queen must be put into a new box of deep frames preferably all drawn and without any brood. If a frame of brood is put in – as has been recommended in some mainstream textbooks – then the bees may well start the whole procedure again because they have been thwarted in their original purpose. They are very sensitive to human interference. If there is some concern about the queen staying in the box, then the two outer frames that are entirely free of brood may be put into the middle of the new box so that the queen is back onto frames that she recognizes to give some immediate security.

Thereafter the supers can be put back above the excluder over the artificial swarm and the hive should continue on the original stand as if nothing had ever happened. If brood is put in with an artificial swarm the beekeeper has lost control and the bees are back in the driving seat. They may settle down or they may swarm again in two or three weeks (most likely). If the queen disappears in an artificial swarm, then this is the time to introduce 2 frames of sealed brood and one queen cell. The parent colony can be reduced to one cell at the same time, so all is not lost. The next hurdle is about whether there are sufficient drones within the locality to mate with the young queens (maybe not).

Unfortunately, as the years go by, we are finding that our bees are determined to be rid of their queen so if an artificial swarm was made she would disappear. Time was that bees were devoted to their queen even at 3 years old and more, but now we are often finding that our bees are dissatisfied with the current year's queen and are trying to replace her by supersedure. The continuity of the colony has become fragile in the extreme. This is a worrying development that threatens our business. As to the cause we do not know, viruses no doubt. The one consistent factor is that over my lifetime pollution has greatly increased to the detriment of all living things.

Nevertheless, there will be areas in the UK where bees still want to keep their queen and an artificial swarm can be made satisfactorily.

I am also mindful that as I wrote in the first article there is a great deal of instability among colonies that have been 'bought in'. So, they are always trying to get rid of their queens. This is a genetic problem. Stability is everything in a honeybee colony and they know it. They need to belong to a person with aptitude, not attitude!

# *June Clover*

I am reading a letter written in 1966 to 'Bee Craft' by R.O.B. Manley saying that he had been keeping bees for nearly 60 years and that his bees had averaged 100lbs per colony in 1961 (1800 colonies). He had little honey in 1962, 1963 and 1965 but the clover failed to yield honey in 1964 when the weather had been better. He put this down to the use of selective herbicides.

This was towards the beginning of chemical farming because selective herbicide was used to kill thistles which also killed the wild white clover that fixed nitrogen to make the grass grow. Thistles could be controlled by cutting them at a certain stage of their development. Large scale beekeeping was then at a great disadvantage and Manley was despondent. We came across this problem four or five years later. Farmers would lose far more money from the reduction in fertility than they lost because of the thistles and nitrogenous fertilizer had to be used which cost money and further depressed the clover.

Very soon all this land was given over to growing grain and around 1970 oilseed rape was introduced which gave beekeeping a much-needed boost but not in the West. The honey smelt of cabbages and set rock hard in the sections making them unsaleable. Many beekeepers packed up then otherwise a great deal of heat had to be used to process rape honey. The crystals melted at 50°C and so a lot of honey was damaged including our own, until we learned to break the crystals mechanically. At one time the wholesale value of bulk honey went down to 50p/lb, until the supermarkets were prepared to buy it. An insecticide called triazophos was sprayed on the rape fields every year, some of it by helicopter on hot sunny days, killing all the bees within a 2 mile radius. Similar chemicals were used in sheep dip, which killed all the fish in our river system. Even in our most remote heather stances, sheep dip was tipped into the burns at the end of the day, all with tacit government approval. The shepherds were soaked to the skin with dilute organophosphorus.

In the 50's agriculture in Northumberland focused on fattening Irish cattle within a 5 year rotation. At that time there were 1100 known beekeepers in the county, all producing clover sections. The cattle came from Eire, having been reared in clover pastures where Irish beekeepers produced well in excess of one million sections per annum. Taylors of Welwyn Garden City who made the sections out of Canadian basswood (Tilia Americana) kept the records.

The general public would eat the finest clover comb honey throughout the land, this was sustainable agriculture. I was delivering honey in Rothbury more recently and a passer-by remarked that honey wasn't what it used to be – a stinging comment but absolutely correct!

We find now, fifty years after its introduction, that nectar bearing has been bred out of most oilseed rape varieties and there is now very little for honeybees in intensive agricultural areas. However, I feel that as commercial beekeepers we have done very well out of oilseed rape, despite the problems mentioned.

There is renewed interest in establishing clover leys but they won't be of much use to bees unless the indigenous variety is planted. I doubt it is possible to buy indigenous seed. Wild white clover needed a night time temperature of 15°C to yield nectar profusely and light rain once a week on thin soils, as this was a shallow-rooted plant. Heavy grazing did not affect the honey flow. There will be places in the UK where this plant still flourishes, Caithness perhaps or parts of Wales.

Children used to pick the flowers to suck the nectar from the nectaries – I remember well!! There was a Scottish Country Dance tune – 'June Clover'. A plant that was revered.

# *Clutching at straws*

I was in conversation with a lady beekeeper from Northern Germany who was telling me that varroa was as bad as ever in her colonies, to the extent she was struggling to keep them going. She was using Buckfast bees bred in Luxembourg and she said that only in the past 2 years had the queen breeders been attempting to introduce genes that would enable the bees to resist varroa. She said that they had treated the bees for far too long. Varroa was found in Germany in 1978, more than 40 years ago. This contrasts with our experience that our black bees are coping with the disease without prompting from ourselves, in only half the time. At least I would like to think that this is the case – I hope so.

More recently I was reading an article written by Bernard Mobus in 1986, entitled 'Clutching at Straws', about a conference held near Vienna which was part of the 1983 Apimondia Congress in Budapest – my parents were there.

Two speakers were Dr Maul from the Kirchhain Institute and Dr Bretschko, a professional beekeeper from Graz. They knew all about varroa before they got it (the bees). Two things emerged. A) that prophylactic treatment would be needed to prevent a catastrophe?? But prophylaxis was not the answer. B) that there would be hidden defence mechanisms within the black bee.

Well we know that there were some feral colonies that survived varroa, albeit very few. Running down prophylactic treatment within black bees is a great risk but that is what has to happen. It has to happen with every type of bee.

# Himalayan Balsam 2021

I was disappointed to read in the Scottish Beekeeper that moves were afoot to introduce a fungus that might kill the river balsam (Impatiens). Most probably the fungus will spread to garden plants and kill them too.

There was a big patch of Himalayan Balsam below Coldstream Bridge and I took 12 nuclei up there to build up for the winter, finding them close to starvation on a return visit I looked and found the Balsam destroyed, most probably by using spray.

We can conclude, therefore, that these plants were the only plants in the locality that were keeping the bees and many other insects alive during August and September. There is a patch of river balsam beside the Chain Bridge that has been there for 20 years and has never expanded because it is bordered by a much bigger, stronger plant with mauve flowers that keeps it in check. I will find the name of this plant in June.

Beekeepers can look forward to farmers in the arable areas growing flowers in order to chase some payments. However, these fields will only last a short time before being ploughed in. But this is better than nothing – give and take, as they say.

I often think that ivy is a wonderful plant, like the Himalayan Balsam it is a robust survivor keeping bees and other insects going with nectar and providing berries for starving thrushes coming from Scandinavia, as well as shelter from the elements for tiny wrens and sanctuary for bigger birds from sparrow hawks. It will also be full of insects during the summer, providing food for birds feeding their young. Ivy honey will appear in your hives as snow white balls the diameter of a cell. Its crystalline structure reflects white light (total internal reflection?), more so than oilseed rape honey that does have a slight colour, so it is easily identified as ivy honey. Supers of honey can be obtained from the ivy in Eire.

Beekeepers will sometimes have noticed crystallized honey in their supers early in the season in areas where there is no oilseed rape. This is most likely to be honey from the wild raspberry, another great plant for bees.

# Beowulf Cooper

I have been reading some work done by Beowulf Cooper in 1967 and which was published by the Village Bee Breeders Assn. At that time there were great numbers of queens being imported from Israel (Buckfast) as well as New Zealand and the USA (6000 p.a.)

As an entomologist he had been called to look at some colonies in an apple orchard near Wisbech. A row of black bees were busy pollinating the apples at 46°F – 48°F (11°C). Another row of colonies with imported queens were staying at home to look after the brood (10 frames of brood). There was a light north-easterly wind at the time, a wind that we all know about, in the East anyway. The imported colonies would also be under great stress finding themselves in an environment that they had never previously encountered. It would take five to ten years for them to adjust or otherwise pass away.

Mr Cooper told the farmer that the colonies with imported queens would come good when the weather improved. He was not pleased about that either because too many apples would be pollinated and the fruit would be small and unsaleable. Only half the apple flowers needed to be pollinated.

Mr Cooper went on to state that drones and queens (black bees) were coming out and making flights on the heather moors in August at temperatures between 46°F - 48°F (11°C). We know about that because many of our colonies superseded at the heather when the old queens were 3 years old and no queen rearing was needed. This was also a way of the bees self-selecting, in that queens that were bred on the moors were cold weather bees, which was ever so important here in the North East. 11°C seems to me to be a very low temperature for queens to be mated, 14°C might be nearer the mark depending on the wind (NW) but it shows how adaptable honeybees are in that every situation presents them with new challenges which they duly overcome by themselves. There was never any problem with queens not lasting 3 or 4 years when they mated on the moors. More recently supersedure on the moors results in queenlessness which is a great disappointment. Queens are often superseded that are 1 year old, which will more than likely be due to a virus.

As time goes on we will have to change every comb in every brood chamber to rid the colonies of agricultural chemicals which will have accumulated over the years, as well as the chemicals we have used to control varroa. Much more space will be needed in the brood chamber to rear drones.

I have a shrewd suspicion, in fact I know, that if we were able to keep bees away from modern agriculture, the queen failure would disappear. My farmer friends tell me that more chemicals (Glyphosate roundup) are being used, a situation which doesn't please them going forward. A treadmill in fact and a costly one.

Agriculture and horticulture have changed out of recognition to become an industry entirely dependant on chemicals other than organic. This spells trouble for bees and beekeepers. It is no good saying that this or that chemical is OK, they are all poisons and all cumulative. What stupidity.

Beowulf Cooper would have no financial interest in promoting the black bees. He was very well respected in his day.

# *Further concerns 2018*

After a very difficult start to the season we have had, like many others, a good crop of flower honey half of which was comb honey due to the very heavy honey flows during June and July. Any hives taken to the heather after the 1st of August got very little such was the severity of the drought. Since then, colonies have stayed very strong going into winter and will need watching for starvation.

Premature queen failure has not been a problem except in odd apiaries. It will be interesting to see how queens mated in 2018 fare next year, hopefully not as badly as those raised in 2017. I know the problem is a disease. Two amateur beekeepers called here during the summer who were very disappointed to have lost all their colonies during this summer due to queenlessness. I might have thought that their problems were due to incompetence, but I sympathized with them and told them that I know the problem and did not know the cause. All that can be done is raise more nuclei and make sure there are vast numbers of drones in the apiary and hope that given time the bees will get over these serious difficulties. We have seen paralysis again this year but in very few colonies. There are some parts of the country that are free of these ailments for now anyway and in our case only a few apiaries are affected.

I see in the Scottish Beekeepers a description of a Canadian commercial operation where oxytetracycline was being administered prophylactically which was no doubt lawful. Now I know very little about this drug, but I was told by father that bees that were thus treated would eventually become susceptible to the disease that was being eradicated and that the drug would be found in the honey.

And so, by the same token we must always keep a few varroa in our hives so that the bees will eventually be able to tolerate them for the long term. At present we are using oxalic acid to hopefully control varroa. I wish I had never used any of the "harder" chemicals that were and still are on the market. Commercial beekeeping was always risky, and the risks are much greater now. But there is always a future to look forward to.

## Beekeeping notes for September 2019

Do you think it would be possible to provide a list of rape varieties that still yield nectar. We are finding that most fields of rape aren't much use for beekeeping and also every year fewer and fewer beanfields yield nectar. So, we are going to have to choose our sites much more carefully with fewer colonies. We first took bees to rape in 1971. The honey was not nice and smelt of cabbages and set like rock.

We have some honey this year, much of it is unsealed, but all in all things are still difficult. There is increasing evidence of paralysis but perhaps the biggest problem was a week's extremely cold weather in May and June, I can't remember when, which set the bees and plants back equally, especially the heather. In some areas the heather beetle has done a great deal of damage due one assumes to the mild winter. The heather came out on 20th August, some of it didn't come out at all on the north-facing slopes as a result of persistent depressions coming in from the Atlantic. This happened in 1985 and 86 and 87 and together with a particularly difficult winter in 1985 just about brought beekeeping to a halt in the north, some honey farms losing every colony.

A gamekeeper who keeps bees on the heather near Duns told us that there were only 3 days when the heather was dusting when the normal span is 14 days. Dusting is when the pollen rises in clouds when the plants are disturbed and coincides with nectar secretion. Most of the heather honey has gone into the brood chambers as one might expect at such a late date and very little brood will have been reared as a result. This might pose a problem later in the winter.

Fortunately, at one of our major sites in the Lammermuirs the gamekeepers planted a good acreage of flowers to provide shelter and seed for partridges and the bees were kept going as a result and got some exquisite honey. Next year we will look around and find fields of phacelia that are being grown by farmers in order to keep the colonies going through a rough spell. All in all, I think the outcome this year could have been worse, much worse like 1985.

We are pressing ahead with our black bee queen rearing programme with Michael Collier and this enterprise will gather momentum as the years go by. I have a single hive of black bees in a garden in Gattonside (the "Gardenside" of Melrose) and it has filled four supers, choc-a-bloc as they say, with honey from garden flowers. Michael Collier has a similar story in Staffordshire with one colony of our black bees in a garden with 150lb on it. So black bees will get the honey if they get the chance. Nutrition from many different sources is perhaps the key to 80%

of colony health. A young queen every year might help with the remaining 20%.

I went to see Andrew Scobbie Snr. at Kirkcaldy in the Kingdom of Fife recently. He is 86 and keeps good health. He was a most outstanding queen rearer of the very best black bees as I saw when I visited years ago. Row after row of queen cells with a queen in very nearly everyone. Black bees are reluctant to build cells in variable weather. They are very cautious. When I told him about the bees crawling as a result of contracting paralysis, he could remember his bees crawling down the garden path because they had the Isle of Wight disease. I'm guessing this was during the Second World War. I know he had bees when he was a schoolboy. The Isle of Wight disease disappeared in an instant relatively speaking. One year it was there then the next year it was gone.

This was helped by the importation of Dutch skep bees presumably just before and just after the war. They were resistant to this most awful disease.

I was giving a lecture at Llangollen about, among other things, total natural resistance being a cornerstone of any beekeeping enterprise and mentioned the Dutch skep bees. A lady from Holland stood up in the audience and said she knew who the beekeepers were that were importing the bees into England all these years ago.

In order for skep beekeeping to succeed the bees had to be inveterate swarmers. So, the beekeepers that got these bees would have plenty problems with swarming but at least they were alive and well.

Talking of skep beekeeping there was a schoolmaster in Pickering in North Yorkshire called Austen Hyde who was a beekeeper. Skep beekeeping was still practiced in places like Rosedale and Farndale north of Pickering where the heather was magnificent. When the time came for the skeps to be emptied of honey Mr. Hyde provided travelling cages for the surplus colonies and sent them by train to Tom Bradford near Malvern who hived them and fed them up for winter. The following spring, they went to Kent for pollination and were sold there to beekeepers to be distributed round the South of England. I remember this because we knew Tom Bradford very well. He was a giant among beekeepers. Skep beekeeping was a cruel business otherwise the skep bees would have perished. I don't like to think about it.

I can remember my father saying that when he was a child there were 60 skeps on the family smallholding. After tea the teapot was filled with hot water and sugar, and this was poured into tiny troughs that were pushed into the entrances of the skeps. This all sounds a bit impossible, but I never knew my father to make anything up. No doubt he would embellish a story to make it better in the telling like I do from time to time.

No doubt beekeepers of that era would have to find ways of feeding their bees when bad weather intervened as we do today. We are 57° north it must be remembered.

# Robbing

I see an article in the SB that is about honeybees robbing fermented honey and dying as a result. Well, this is really about robbing, a common activity of honeybees, particularly yellow bees.

When bees find an unguarded source of honey they try to clean it up before other colonies get started. In practice every colony in the locality is visiting and they start to fight among themselves in order to take control of this unexpected source and many are killed in the process. Additionally, any humans or other living things that happen to be around at the time will be stung and chased away. All colonies will have many guard bees at the entrance ready to kill any aliens that might come near. This is because some colonies, no doubt very strong bees, will sometimes go berserk and try to gain entry into any colonies that they consider vulnerable, i.e. nuclei or colonies with failing queens. And so mayhem breaks out.

Many years ago some beekeepers were taking off honey at a communal heather site when one assumes they left in a great hurry probably leaving a hive open. The resulting robbing went on for three days with the lady in the nearby cottage unable to go into her garden without being attacked. Modern bee suits have helped, many beginners into beekeeping. It wasn't always like that; old style veils were only of any use when the bees were in a good mood.

And then there is another type of robbing that goes on silently. A strong colony gains access without trouble into a weak colony (generally queenless) and they steadily carry all the honey away without any fuss. Other colonies around about are not involved although they will show some alarm. It is very likely that the robbing colony will be the first to provide a swarm into the empty hive the following spring. They will be taking possession in early March cleaning it out. Conversely about 15 years ago we noticed that colonies that had successfully come through the winter would not rob the heather honey out of hives that have died through the winter. In fact they will not go near them. This is still the case. There must be a residue of some sort, perhaps to do with varroa, which keep the robbers away, either that or the treatment for varroa which might be in the brood comb. A portent of troubles still to come.

A further important consideration that beekeepers must be aware of is that if a super containing any 'loose' honey is put on a colony during manipulations those bees will come out in a few minutes and sting the beekeeper when he or she is going through the next colony. So all wet or extracted supers are better put on the night before and all will be quiet in the morning. This is an elementary precaution because it is so important for beginners to build up confidence. If a honeybee colony thinks that the beekeeper lacks confidence, then they will sting them and make them feel even less confident and so on. So, it is important to go into them on a good day and avoid elementary mistakes like crushing them or dropping a frame. Over confidence is another problem. (too much smoke) Ultimately it should be possible to go through them without a veil or gloves. Old hands often did just that. It is important to build a relationship with your bees.

# *Practical Beekeeping*

**Swarms and casts**

As I write this in early July, we are having some good weather allowing the bees some respite from the very poor June weather (daily temperatures down to 7°C and much lower at night)

A common problem at this time of year, after such a difficult period, is that young queens will be flying every day. If the beekeeper is working through colonies young queens will alight and go into supers that are off the hive and become trapped there above the queen excluder, once the hive is reassembled. She will then become a drone breeder in the super.

Similarly, if a colony is found to be full of hatching queen cells all the queens should be released (and I mean all of them) and there will be only one queen left in the morning. If a queen cell is missed, then the colony will throw a big cast in the morning and casts are notoriously difficult to keep in an empty hive. They would rather travel far and wide because the virgin queen is able to fly with them. This is to do with the bees breaking new ground. Swarms are likely to go into established sites close by.

To get back to the original discussion if there is the slightest suspicion that virgin queens may be flying during manipulations then a super must be twisted so as to give an entrance above the excluder. I continuously forget to do this with the result that there is a drone layer in the super. Twice recently I have found virgin queens in the cab of my truck parked in front of the hives. (wrong place)

Black bees and maybe other varieties as well are able to prevent virgin queens hatching during bad weather for obvious reasons. They let them all out on the first good day so they can throw a cast or two. This is another example of their sophistication.

If a person phones in May and tells me they have a swarm on their property I tell them that a beehive is close by. If they phone up with the same story in July the bees might have come from three miles away. A prime swarm generally settles low down on a branch whereas a cast settle high up in a tree because the virgin queen can fly just as well as the workers (repletes

A swarm or prime swarm is accompanied by the old queen in the hive. A cast is accompanied generally by a number of virgin queens. This accounts for the loose nature of the cast (the cluster). They don't adopt one queen until they have settled in their new abode which may be several days/miles away. Sometimes they die of starvation in the process. A tiny cast has only one queen.

In more recent times the queen in a prime swarm may fail as soon as she is in her new hive (at one year old). She has stopped egg laying in order to be light enough to make the journey and many land on the ground because the queen can fly no further. The bees stay with her because they are not orientated on the old hive and, unless found by the beekeeper, they die of exposure.

However I digress again. If the old queen in the swarm fails immediately then the new colony is doomed. If she lays a few eggs then emergency cells will appear otherwise a queen cell will be needed or a frame of brood once the old queen has been disposed of and they might survive.

The problem of queens failing before they are a year bold represents a major threat to beekeeping in the UK. We have, and have had, for four or five years now big successful colonies finding themselves without a laying queen in mid-July as opposed to a break in egg laying. The queen has remained in the hive in a sterile state whilst still giving off the pheromone that binds them together (9 oxydecenoic acid) or so I think. And so the colony drifts into a state of queenlessness. The bees become so demoralised that they aren't interested in a frame of eggs, either that or they still think they have a queen. A ripe queen cell may be accepted otherwise a nucleus will be needed. All too often they are cleared out by wasps (great big colonies)

We have hundreds of nuclei going at this time of year and I notice that despite having perfect brood patterns some of them are already building supersedure cells to replace queens that are only a few weeks old.

As to the cause of this chronic problem we have no idea. From chemicals perhaps or some exotic virus affecting the drones or a combination of factors I see articles in the beekeeping press by learned individuals on subjects that are totally irrelevant to practical beekeeping. We live in an age of cluelessness. Even if we did know what was the matter with them there is little we can do apart from making nucs. The French beekeepers told me this years ago.

We must remind ourselves that honeybee colonies organise themselves to be immortal, otherwise they would be like wasps and bumble bees and die out every winter.

If it weren't for beekeepers there would be no bees. Going back to the 1950's imported queens generated a good deal of trouble because they weren't acclimatised and became susceptible to every disease. More recently imported queens have kept the industry going although I wonder how many hundreds of tons of sugar it has taken to maintain these big colonies this year. And then they are susceptible to these modern diseases as well, including paralysis which is becoming a problem in the north. I can see that paralysis is to do with malnutrition more specifically the lack of certain essential constituents. Previously black bees were highly resistant to this problem. In the north they would need to be. In intensive agricultural areas even more so. The tweed Valley where I live is reputed to be one of the most productive agricultural areas in the country. And yet there is no field, other than organic, within this vast area that would produce anything at all not even a weed if it wasn't getting the full measure of fertiliser and chemical. And I am not alone in thinking his, many farmers also.

I enclose a letter from a much- respected retired bee farmer in the north of Scotland which I think is very well written

I must apologise for the rambling nature of this article

### *Letter from bee farmer in the north of Scotland*

West coast varroa – first found August 2016. Count to October 2018 very low and not treated, mites with legs missing.

East coast varroa - present since 1999 and treated twice per year until 2015 then once per year. Treated after first honey flow (formic acid). Very low count October 2018.

It seems to me that the bees are beginning to give us a help in combatting the mite. I hope so then we will be able to much reduce the amount of chemical warfare in the battle.

### **Queen mating and laying period.**

As I explained some time ago queens bred on the west seem to exceed by far those reared on the east, laying for at least three to four seasons if allowed (I have brought 2year olds back to the home apiary here on the east to counter emergencies and they have managed another year).

Here on the east we experience supersedure, sometimes within a year, suggesting that the drones are unable to cut the mustard.

Where drones in days gone by used to congregate we now have a quarry which may have some detrimental effect but that surely cannot be the reason. I firmly believe that it is in the soil. Some say that it is caused by the effect of the varroa mite but that in itself cannot be the problem as we have varroa now on the west coast.

The few farmers that I have been able to approach on the subject of insecticides inform me that for quite some time here on the eastern side of the country (Easter Ross and the Black Isle) neonicotinoids were used right up to being banned and several others such as Lindane.

Although the use of these are banned, we are left with the question of how long does it take for the effect of these to completely leave the soil. For example, could it be that there still could be miniscule traces in the soil after 20 years? Combine that with traces of other chemicals from more recent years and we could have a deadly cocktail on our hands.

This cocktail will leach through to the headlands and hedge rows thus affecting all other wayside plants and whilst one plant with, say, chemical A may give pollen and nectar to the bees, a plant with chemical B doing likewise is now beginning to build up the cocktail in the hive. This surely must have some detrimental effect on the bees' food and indeed to the honey we harvest.

We are certainly living in dangerous times, and I just hope that no one starts spreading chemicals over the west coast or we will lose a good many good acres (square miles) of excellent bee grazing.

**Note:** - If in doubt as to your ability to cope with all this advice then make up a 4-frame nucleus (2brood, 2 honey) with one queen cell and leave it on the original site with the flying bees. Then move the original brood box four or five yards away. Very soon they will have sorted themselves out. Remember that both colonies may need to be fed. This is a practical solution to swarming problems to enable the beekeeper to go home.

# *Hives cannot function without a queen*

The biggest problem with the bees is that the queens are not living more that a year and hives are suddenly left queenless at any time of the year. Even if the queen dies in the Summer the bees often make little attempt to replace her. And so, their basic instincts are affected. Fifty years ago or less queens occasionally survived into their fifth year. Hives sometimes went queenless but this was not seen as a problem. By and large the bees survived and flourished. Now there are no feral colonies, not for a period of fifteen years or so. That feral (wild) colonies are unable to survive tell us all we need to know about the plight of honeybees. And then we are reliably told that 50% of drone bees are infertile for reasons unknown and so queens are not properly mated. When there were plenty of feral colonies there were so many drones flying that every queen got mated.

And so, the bees have endured one big epidemic (varroa) and just about got over it (there have been epidemics in the distant past). Unfortunately, in their weakened state other diseases have appeared that come and go and then we have queenlessness. Farm chemicals must take some of the blame because they are found in hives and honeybees are extremely sensitive to chemicals. They are held together as a colony by a pheromone exuded from a gland in the queen's head (9 oxydecenoic acid). Honeybees can detect a source of nectar 4 miles away. They were once the great survivors, now they face an uncertain future. Farm chemicals are like a sledgehammer to them (fungicide/cocktails?). However, there would be no crops without chemicals and cheap food is a government priority.

Asking around the situation is much the same throughout the country, perhaps worse in France and much worse in America presumably because they are industrial beekeepers and use more farm chemicals. Even in New Zealand there are similar troubles, and they may not have so many chemicals on their farmland, if any. And so, nobody knows what the problems are and we can only guess.

As a partial remedy it will be necessary to change the queens every year weather permitting and provide a much greater number of drones in hives by inserting drone brood foundation. Even then losses will be widespread and variable. Sometimes a whole row of hives becomes queenless. So, this is a virus that is contagious? The bees will eventually become resistant to these infections over several decades because they only have a short window (May/June) in which to recover, and they won't do that without a lot of help from the beekeeper.

Even so after a disastrous start to 2018 the bees got a lot of honey in July. Little heather honey was gathered because of drought. As a result of a mild Winter in 2019 they have gathered a good crop of early honey.

They are from time to time on the verge of bouncing back. However, there are very, very few feral colonies that survive and that is the ultimate test of their health. For practical purposes their health status stays resolutely at rock bottom. This is a shame because they are wonderfully sophisticated and industrious. I am referring throughout to indigenous black bees.

# *Insecticide in London Parks*

I was surprised to see an article in the Scottish Beekeeper casting doubt on the findings of a study about the gradual disappearance of insects in Germany. It has long been a talking point in these parts about the total disappearance of flying insects from the front of cars over recent years. At one time it was difficult to remove crushed insects from shiny paintwork but not any longer because they are not there. And so I would conclude that the Germans got it about right even if their procedures were open to question.

I saw an article in the Telegraph recently by two Cambridge scientists telling us that there were too many honeybee colonies in London and that this fact was driving down the fortunes of other insects. If one looks at an aerial photograph of London there is certainly a great amount of habitat for insects and other wildlife and so I would suggest that these scientists were speculating.

Many years ago a lady visited here who said she had a responsible position in the Ministry of Agriculture. She said that systemic insecticides were being incorporated into potting compost used in the parks and gardens in London and in her opinion this was not a good idea and that there were no regulations governing the use of these chemicals in horticulture.

Well, there you go, it would not surprise me if the same chemicals were used today in order to keep flower beds insect free. I read the other day that four agricultural chemicals had been withdrawn because they had been found in drinking water.

A friend who makes high quality bread locally noticed a problem with the dough and eventually discovered that the flour he was using contained glyphosate. And now I read that 50% of honey tested (80% in the USA) contains pesticide including neonicotinoids. A beehive is a receptacle after all so this is not surprising. I am extremely concerned about the present state of beekeeping with all the queenlessness and loss of instinct and ability. I said some while ago that black bees would go on indefinitely, keeping themselves and giving us a living but not now. They might hopefully get over varroa and the viruses but they will never cope with agricultural chemicals, never ever.

I have been told on some authority that fungicides have been found in royal jelly and this will undoubtedly affect the development of the queens and shorten their lives. We have noticed that queens that have been laying for a few weeks are suddenly dying/disappearing and the bees are making little effort to replace them. The queens will have to be fed large amounts of high quality food daily just to keep up with egg-laying. It is certain that minute amounts of pesticide will be present in that food thus affecting their health. All this is conjecture but maybe I am getting close to the point.

However I still think that viruses are the main problem and they have got going as a result of stress brought about by varroa as well as variable levels of malnutrition caused by too many colonies being sited in areas where there are too few plants to sustain them especially during poor weather. At one time black bees would stand any amount of malnutrition and come up trumps. The agri chemical problems are just another dimension to a poor situation. A combination of adverse factors drawing the bees ever downwards.

# Concerns about the health of our bees

About 10 or 12 years ago I noticed the bees were not going about their swarming procedure correctly and that they had lost some of their instinct. Swarms were flying around with queens that were not viable and hives that had succumbed to varroa were not being filled up as one would expect.

More recently queens, including young queens, have been disappearing during the summer, never mind the winter and bees have made no attempt to replace them. This sort of behaviour indicates that the bees have lost their will to live. Years ago honeybees often stored enough honey to last them for 10 years and they kept queens that lasted for 5 years, indicating that the hives were entirely free of chemicals or disease. This underlined their determination to survive as befits a highly sophisticated social organisation. They were a powerful force in the land beholden to no-one.

Nowadays I am afraid they are on the retreat and even beekeepers like ourselves find it difficult to keep them going. We are going to have to keep one small colony for every productive colony and this means industrial beekeeping which we have always avoided.

A correspondent in France who would know what he was talking about (Thierry Fedon) is saying that the fertility of queens and drones (according to research) was down by 50% and practical experience would suggest that this was correct.

As for reasons I suspect that viruses have got going on the back of the Varroa epidemic and spread within an apiary resulting in heavy losses whereas other apiaries are unaffected. Nutrition which is beyond the beekeeper's control has a part to play as well as the lack of shelter.

The French and Germans blame the chemical companies square on and there is no doubt that beehives are a receptacle for chemicals that are carried in. Humans and every other living thing the same. Chemicals that weren't there 60 years ago. My perception of agricultural chemicals over a long period is that they were constantly being taken off the market because they were found to be dangerous (lethal) after a period of use which means that the vetting processes were useless. And maybe still useless. Constant use of a particular chemical over a long period is bound to generate problems whereas occasional use might not.

And then we hear of research in Germany that tells us that perhaps 80% of the flies which we used to see stuck to the front of cars have disappeared over a period of 20 years. Well we have noticed that too. Whenever we hear of some important research within our subject, we subconsciously remind ourselves that yes, we have noticed that too. And so, the future for beekeeping worldwide and particularly for high latitudes is not assured. Far from it. Feral colonies are still very few and far between. The number of feral (wild) colonies within a locality would be an extremely accurate indicator of honeybee health. At one time there were many more feral colonies than there were domesticated ones. Now there are none that could be considered permanent.

*Willie Robson*

# Observations from Willie Robson

**2015 Beekeeping Season**

Bees came through the winter in very good condition as a result of a good summer in 2014 and a very warm autumn. It was noted that bees were gathering honey from the ivy on November 15th 2014. This is very unusual. The bees did well on the blackthorn and other prunus varieties in March/April 2015 but then we had an extremely cold period during April/May. The bees came through this period easily because they were in good heart. Some honey was gathered from the oil seed rape and subsequently a good amount from field beans. However during July the weather became desperately poor with 20 days of rain and night temperatures down to 5 degrees. Most plants cease to yield nectar under such circumstances and the bees cannot maintain their broodnests without eating up all their honey which they did. We then had to feed them all in a big hurry until the middle of August. Fortunately the wind went into the south and brought warmer air from the continent and because there was plenty moisture at the roots of the heather a good amount of honey was gathered in late August. This has left the bees and ourselves in good stead for another year and so we have been very lucky especially as nationwide very little honey has been gathered.

The general consensus is that throughout Europe beekeeping becomes a little more difficult every year because of premature queen failure. Reasons for this includes stress caused by the varroa parasite or stress caused by the use of chemicals to control varroa or the presence of agricultural insecticides in the system and, in this country, variable weather. Most likely all four.

**Our colonies this spring**

We are more than surprised at the condition of our honeybee colonies. Nearly all of them have come out of the winter as strong as they went in which, considering the weather that we experienced last July, is quite remarkable. The wind went into the south during August 2015 and gradually the bees recovered their condition and as there has been very little bad weather since they have maintained their condition to the extent that we cannot recall seeing bees so far forward at this time of year. While there is plenty pollen about for them they are using a great amount of honey in order to maintain themselves and their brood nests. Sometimes the blackthorn keeps them going but temperatures are a bit low for this plant to yield nectar. So beekeepers everywhere will need to be extra vigilant and take care that the bees do not starve after such an easy winter.

**Further thoughts on the 2015 season**

I have seen the odd queen wasp about this spring (2016) but not many. I remember last year during the atrocious weather seeing two wasps' nests where the adult wasps were tumbling out of their bykes due to starvation. It was also reported in the Daily Telegraph that wasps were dying of starvation. I cannot recall this happening ever before although one would not notice unless by chance. I suspect that during the summer of 1985 that this might have happened but one thinks of wasps as being exceptionally resourceful which they are but if the weather is bad enough they go too. Unlike the honeybee they have no stores to fall back on. Wasps are omnivorous but mainly carnivorous (flies, caterpillars, green fly etc) i.e. beneficial.

Last year, 2015, during the bad weather we found many colonies that were queenless and, in normal fashion, we put a comb of eggs from another colony in the brood nest for them to try again. But then, subsequently, we found that these colonies had reared the eggs up into worker bees but some had not made any attempt to rear a queen, presumably because of continuing bad weather and very low morale. A local queen rearer told me a similar story where his bees (Buckfast) refused to build any more cells. All the bees in the area have made the decision that queen rearing is no longer a viable proposition and even if the colony is queenless they will not make any attempt to get going again.

These queenless hives were soon empty but no doubt this spring they will be full again because our bees are in such good condition and will swarm immediately into the empty boxes. It is as if they know that by waiting for another season the colony numbers will get back to a full count

Indigenous honeybee colonies support one another by providing drones in order to promote close breeding but hopefully not inbreeding and also make up neighbourhood losses as quickly as weather and nutrition will permit. They would like to be immortal so that nobody would notice that they had died out and got going again.

These facts underline the one major problem that they have not mastered during evolution. If the queen fails then the colony is very often lost. This situation is highlighted by premature queen failure which is becoming very common.

I am told that this is being caused by a virus and that beekeepers have to requeen annually and that in California queens are barely lasting 6 months. This year, 2016, we intend to flood our apiaries with drones but I don't suppose it will make much difference. I think I mentioned in a previous article that marked breeder queens in the North West of Scotland where there is no varroa were lasting for 4 years. I believe4 it. A good beekeeper needs plenty overwintered nuclei to prevent colony numbers dwindling. Polystyrene nucleus boxes would be a great help in this respect.

## Honeybees as pollinators

I was very surprised today, the 27th April on arriving at an apiary at Fenwick Steads to discover that bees were working with a purpose at a temperature of 6.5°C (45°F)

Previously my father had said that bees would work well at 50°F so it was interesting to see how much further down the temperature scale they would go. A field of oil seed rape was in flower about half a mile distant and it was sunny. The sun would put some warmth onto their backs and this, together with the fact that they were black bees and no doubt able to absorb a little more of the sun's heat into their bodies, would enable them to battle their way through the bitterly cold wind. If it had been overcast they couldn't have done it. The colonies were prosperous and needed supers.

I continued to another field at New Haggerston and there the bees were dormant and not prosperous although they were in a field of rape that was in flower.

And so we can conclude that the first field of rape was perhaps an old variety that yielded nectar at very low temperatures and the second field would be a newer hybrid variety that perhaps didn't yield much nectar at all. We shall see. It also shows that honeybees are exceptional pollinators when there are no other pollinators about. I sometimes hear that there are more efficient pollinators than honeybees but there aren't many that make themselves available over such a long season and in such numbers and at such low temperatures. I remember that on the 18th November 2014 the bees were working hard on the ivy, albeit at a fairly high temperature. Ivy is a very useful plant giving food and shelter from the weather and predators to many wild birds.

And so it is not possible to overstate the honeybees' importance as pollinators of both crops and wild flowers and so beekeepers are important too but we never much think about it until we see the bees working at 6.5°C (There is a thermometer in the van!)

## Oil Seed Rape Honey

It must be more than 40 years ago that farmers in the UK started to grow oil seed rape. I remember that the nectar smelled strongly of cabbages but to those of us that were dependent on our bees for a living this was a useful development especially as the indigenous white clover that previously sustained our colonies was fast disappearing from the pastures.

However, very many local beekeepers packed up for good at that time and skills were lost. Section honey was unsaleable because the sections were rock hard. Extracted honey went rock hard in the jars and bent the teaspoons that were forced into it. Rape honey got a bad name with the beekeepers and the customers and beekeeping generally took a nose dive.

The purpose of my article today is to look again at the methods available to beekeepers to deal with this very useful crop. Most beekeepers will try to extract rape honey before it granulates and rape honey today is slower to granulate than previously particularly in new comb. So this is a good time to draw new foundation. Points to watch are as follows. If nectar is raining out of the combs then they must not be extracted or the honey will ferment. If just a few drops of nectar are seen dropping out of the corner of the comb then it is worth taking a chance as the moisture content will equalise once the honey is stirred in a pail. If in doubt then borrow or buy a refractometer. The moisture content should be below 18%

When honey is extracted it will contain many tiny bubbles of air. These bubbles will act as a nucleus for tiny crystals to form that are the basis of soft set honey. This honey will look like milk when it sets because the crystals are so minute that they act as prisms and reflect white light back into the eyes of those that look at it. I think that this is called 'total internal reflection'. Ivy honey has the same property as most of you will have noticed. This then is the honey that the beekeeper needs to retain to use as a starter for future batches of creamed honey. If the honey is warmed to strain it the subsequent crystals will be much bigger and the honey less palatable. Skimming the wax from the surface of the pail will be adequate. All rape honey should be sold as soft set honey using the Dyce process. Descriptions of this can be found in good text books or no doubt on the internet. The beekeeper will need a small insulated warm cupboard to do this together with a thermostat (up to 60°C). Never heat honey above 50°C.

If the honey has set in the combs then they must either be cut out or scraped. The supers will need to be kept in a warm room for 48 hours and the honey put into pails making sure the cell walls are thoroughly broken down. It is worth noting here that if a colony has swarmed all the rape honey will set immediately in the hive because there are no longer enough bees to keep it warm and keep it liquid. The pails of honey then need to be put in a cupboard at 45°C. Honey crystals will find their way to the bottom of the pail and can be used for cooking. Don't try to heat these big crystals as the temperature needed to do this will spoil the honey. Clear honey can be poured from the middle of the bucket and strained, creamed and sold. (Dyce process). The wax could usefully spend another day to get the remaining honey out and then soaked in water overnight before rendering otherwise it will be sticky.

I am well aware that there are uncapping trays that will deal with solid rape honey but the honey will be ruined and only fit for cooking. I am also aware that there are some bigger machines imported from Denmark (Dana Api Melter) where perhaps the separation process (wax from crystals) is not so sudden. It is all about the temperature between the heater (water) and the honey. The Danish machine heats the honey, I think, through air like an electric fire inn an enclosed space. This might be a bit better but not much. You cannot cook honey without destroying its nature and the taste. The only thing that the British beekeeper has in his favour to be able to compete with supermarket honey is that he is able to provide honey that tastes of summer and the flowers that the honey has come from.

Consider if you will a barrel of honey that has come from abroad. It has to be heated for a long period to liquefy the contents sufficiently to make it flow through a heat exchanger in order to heat it to 170°F (76.6°C) so it can pass through a fine filter and kill all the yeast present and remove just about everything else as well. Even heather honey gets this treatment. If heather honey is heated above 50°C it is ruined. Light and time will degrade any honey and in particular heather honey, even in the comb. The reasons behind this would take another article. Honey in pails should always be stored in the dark.

When a lady phones up for raw honey we may feel that she is a bit of a nuisance but actually she has got it right. Unfortunately the terms "raw honey" and "cold-pressed honey" are too often used by shady honey packers. I hear sometimes of heather honey being sold in a high class supermarket in our capital city where the label states that the honey has a hint of caramel. Enough said.

As part of an integrated management system nearly all our honey is produced in shallow frames fitted with only 1" of foundation. Most of the rape honey is extracted at very low speed in a large extractor. Some of the rape honey is left on the hives to prevent the bees starving if the weather turns bad in June. When this honey is brought in during July the comb is cut out and broken up in an industrial potato masher. It is then stored in a warm space for a day or two to allow the wax to rise to the surface. The honey which still contains a lot of wax is pumped through a big slow speed centrifuge that divides the mix into three phases 1 Heavy crystals 2, liquid honey and 3 wax flakes. The heavy crystals and honey are then pumped through a very old homogenizer. The big crystals are pushed through a serrated cone sitting in a hollow and backed up by a spring set at 3000lb per squ. In. This breaks the crystals up without having to heat them. Thereafter the honey is stirred and put way to be bottled at a later date. Thus we are able to produce a soft set honey directly from rock hard comb using only background heat, certainly nothing above 40°C as well as minimal filtration.

When I was giving a lecture in Sweden about 20 years ago I was complaining about the difficulty of separating wax from honey that was mainly crystals. A gentleman in the audience, an amateur beekeeper from Norway, told me afterwards that he knew how to make such a machine that would do the work. And he did. And we have used it ever since. Which just proves once again that when lecturers such as myself are out and about there will always be someone in the audience that will have some useful information.

When we first began to break up rape honey crystals we had a big granite roller mill that had been used to make nougat in France. A beekeeper friend made a beautiful set of stainless rolls for it (36" by 12"). Unfortunately this machine pulled air into the honey and made it difficult to sell. The rolls started to wear fairly quickly as well showing that rape honey crystals were as hard as stainless steel.

In conclusion I feel that most beekeepers will have found ways of dealing with oil seed rape honey but this article might be useful for beginners. It is most important to use as little heat as possible so that the finished product tastes like honey. And well prepared rape honey is a superb product and greatly in demand. 40 years ago it caused many beekeepers to pack up for good.

# The Smith Hive

The Smith hive was designed and built by W.W. Smith of Innerleithen, Peeblesshire in the Scottish Borders. He kept 120 very powerful colonies of black bees for heather honey production. He read many books about beekeeping written by American authors and decided to build a hive on the American pattern but to the British Standard size. The dimensions were 18 1/4 by 16 3/8 in. to accommodate 11 BS deep frames (short lugs 15 1/2 in.) using 7/8th in. timber all round with a frame spacing of 1 15/32 in. Inch and a half was too tight. The brood bodies were made of four pieces of wood half lapped at the joints and were cheap and simple to make – much more so than the National.

The floors could be extended at the front or otherwise flush down the front and were 3/8ths deep to prevent mice getting in (full width entrance). The crown boards were flat on the bottom side and with a shallow rim on top (top bee space) and often incorporated a swivel to provide an entrance for a brood body containing a nucleus on top. A feed hole was provided with a cover. The roof was constructed to fit over the brood body with suitable clearance all round and could be up to 10in. deep and covered in flat galvanised iron or aluminium folded at the corners. The shallow boxes were 5 7/8 in deep, again made from 7/8th in. timber and would usefully accommodate 10 Manley-type shallows 1 5/8th in. spacing for honey production.

R.O.B. Manley designed frames with closed ends right to the bottom bar that didn't swing about in transit. They also double-walled the super on the two long sides which was important on heather sites. The hive would normally be constructed in Western Red Cedar or in our case, *Thuja plicata*. Willie Smith built his hives in white pine which, once it was dry, was very stable. He also painted the hives in white lead paint which looked well. This was a cheap and simple hive to make in large quantities. The worst feature of the Smith hive was the handholds which were very often too shallow to get a grip.

Willie Smith was well known to our family, and we learnt a great deal from him. He built his colonies up by giving the bees several deep boxes from which heather honey had been scraped and which still contained a good amount of honey. Thus, they became very strong to be ready for the bell heather in July and the ling in August. The bees stayed in the same sites winter and summer and were able to fly to the heather without being moved. This was a very late area and most of the queens were superseded during August, sometimes when they had reached four years old. Mr. Smith ran a very successful business.

He kept the bee space on top according to the American pattern so he could tip up adjoining brood boxes without dislodging the frames in the bottom box because of brace comb. This might not be the case with a bottom bee space.

**The Cottage Hive**

The cottage hive was favoured by Northern beekeepers because it was small and not so cumbersome as a W.B.C. and so could be used in migratory beekeeping. The brood box was designed to accommodate eight or nine BS frames with long lug (17in) and metal ends and was single-walled with a fixed floor and a very small entrance – 6in. – together with a porch to deflect wind and rain. Above the brood chamber the hive became double-walled, the "lift" being the same exterior size as the brood chamber but having a rim around the lower edge so that it could not be knocked off by a sheep. Inside this lift, or outer cover, there was room for 21 sections packed all around by discarded clothing. The roof was gabled and made entirely of wood and again the roof was a tight fit over the outside of the lift so that it was not easy to dislodge. These hives were also painted in white lead paint and they were generally made by the village joiner. Handles were fitted to either side. Sections were 41/4 in. square and were solely for comb honey production, mainly wild white clover and ling heather. Indigenous white clover was the mainstay of British agriculture in the early part of the 20th century and produced, under the right conditions, vast amounts of nectar.

There was also a Glen hive which was used in Aberdeenshire which was similar to a cottage hive but contained 14 deep frames which was very wide. I cannot think why this was done as bees like to move upwards especially in cold climates.

# Black Bees – Decisions critical to their survival

In my recent article about the black bees I talked about thrift. After the longest day the bees begin to think about winter and the first thing they generally do is put the queen off the lay, i.e. no more eggs for a good while. This is so that if in ten days a honey flow should arise all the bees in the hive that are able to fly will go out and forage and none will be required to stay at home to feed and care for the brood. The running costs of the hive are reduced to nil and all the honey can be stored for winter. Ripening honey goes on through the night and intensity of the honey flow can be judged by the noise coming from the entrance. Now it may be that the honey flow may never come but at least the black bees have prepared for it. Hybrid imported colonies may keep breeding right through and in the event of an intermittent honey flow, will end up with nothing and have to be fed. In this area of North East Britain, honey flows are few are far between, particularly in an area of intensive agriculture. The climate is maritime (low atmospheric pressure), as it is in Wales and Devon & Cornwall. And yet the black bees will get some honey in a very difficult year because they put their queen off the lay at the critical time. This is their way of keeping a small brood nest and getting some surplus honey. In a heavy honey flow the hybrid colonies will get half as much again because they have such huge populations but only if plants are producing plenty of nectar.

During high atmospheric pressure in July (a heatwave) many different plants come into production and it is not possible to tell where the nectar is coming from. These plants only yield nectar in ideal conditions and only in certain years. So for all we talk about looking after bees, in a correct manner it is wild plants that dictate the final outcome – plants are not machines.

During settled weather the bees can fly 3 miles (20 sq miles) and as time goes on so the opportunity to gather nectar increases greatly with every bee that can fly working at full capacity. At the end of the honey flow at least 60% of the bees will die within a week and the hive will be depleted as if it had swarmed. In the North this may only happen two years in five (if we are lucky)

Thereafter the queen will be brought back onto the lay and in 3 weeks the hive will be back to full strength preparing for winter. Unfortunately, in recent times some queens fail to come back into lay and the hive becomes queenless, often with 100lbs of honey in it which is duly robbed out. A huge loss to a commercial outfit. We see our colonies coping admirably with varroa, only to be faced with excessive levels of premature queenlessness. A worldwide problem with no answers, but not everywhere. The time was that honeybees were able to look after themselves but at present if indigenous colonies are to survive in any numbers, beekeepers will have to work assiduously to keep them alive. Things have come full circle.

# Viruses, Varroa and Dilution

I noticed two articles in the Scottish Beekeeper that would be worthy of further comment.

Nigel Hurst was commenting on a swarm emerging that subsequently could not make up its mind about the next step. This behaviour may be caused by a virus within the colonies that interferes with the bees' instincts in that they don't know what they are supposed to be doing. Additionally, the old queen is likely to be faulty due to unsatisfactory mating the previous year, which makes the bees uncertain about what to do for the best. It may well turn out that the young queen will not get mated either, leading to colony collapse (robbed out) during the summer. This is common as these viruses can occur right through an apiary leading to the loss of several colonies.

We are advised that every last varroa must be eliminated from the colonies during the broodless period and that apiaries must be flooded with drones in the hope that some may be fertile at the critical time.

Previously we were trying to keep bees that were able to cope with varroa by reducing the treatments, a policy that looked as if it was working but the continued presence of viruses has caused us to change our strategy, which is a pity because the bees will never become fully resistant to varroa. I remember that about 1960 in response to the presence of acarine and nosema, no treatment was given and colonies that were susceptible to these diseases were allowed to die in the winter and those that were left were resistant, although the diseases could still be found in 'healthy' colonies. This is not happening with varroa (viruses), at least with us it isn't. There are other beekeepers further south who claim to have varroa resistant colonies and I don't disbelieve them. The problem with our lot is 'dilution' together with other factors, namely occasional malnutrition, transport and weather. We are a commercial outfit and have to make the best of a difficult job.

Dilution means that too many colonies within an apiary or apiaries have little resistance to varroa and cause the resistant colonies to lose some or all their resistance by outcrossing. This situation could be resolved by line breeding including artificial insemination. A huge task but possible. We rarely see varroa in our colonies.

When discussing 'dilution' I am reminded that we have taken great care never to upset our black bees through rough and untimely handling. When we come across an aggressive colony (perhaps three in a hundred) we just ignore them and move on knowing that in a year or two they will return to being docile by 'dilution' with surrounding colonies that are all beekeeper friendly. Similarly black bees will throw off an invasion of imported strains in a very short time (ten years) by dilution and other more complex strategies because there are too many robust indigenous colonies within the area. In England the reverse has happened and imported strains generally predominate.

Some five years ago I put a colony in a well-kept private garden because the owner wished to watch them 'working'. During this time the bees never stung anyone. They would see and recognise the owner every day as well as others that they would also recognise. I could work on them without a veil, and it was a huge colony. This year a cast came into an empty hive that I had prepared alongside and when I go to the big hive this little lot are out and looking for trouble as soon as they sense my footsteps. This tells me that the previous owner of those bees had a poor relationship with them in that they were ready for trouble even as he or she approached and so I will have to take that colony away and put it in with others that know how to behave until it loses its bad temper by 'dilution' (assimilation).

This is an object lesson to all beekeepers in that if you keep black bees or bees of a mixed race try never to get on the wrong side of them as they will remember right through to the next Summer and beyond because they have a corporate memory. They are also liable to sting all the wrong people.

# An Assessment of the Black Bee going back 100 years

Originally in skeps (our family had 60) and then into cottage hives (north) and W.B.C (south).

- Swapping sites with feral bees and often mated with feral drones.
- Very small broodnests (6 frames of brood).
- Kept for sections (comb honey) which they hated because they were designed by humans.
- Prone to swarm every year.

Often got good crops of honey, plenty flowers and many acres of wild white clover (indigenous). This was the most important plant in agriculture. Every farm (subsistence agriculture) and village kept hives of black bees, more hives sometimes than villagers.

There would be many imports of bees before the 1st World War and thereafter of exotic types but mainly Italian. Italian queens went to America also to become Starlines and others (hybrids). I think all the big honey farms in the south would see Italian queens apart from Brother Adam. They would be much easier to look after in large numbers and get more honey.

At this time the Isle of Wight disease was raging. Whole apiaries were lost and beekeepers were social distancing because the bees were kept in cottage gardens (among the hollyhocks)! They could not have been defensive. Many beekeepers would have gone through them without a veil.

These losses were partially made up by imports of Dutch Skep bees that were resistant to this most virulent disease. These bees would not align themselves with any progressive management. A step back no doubt, but at least they were alive. Then one day this disease just disappeared according to Andrew Scobbie who kept bees just after the war.

Sugar was given to beekeepers during and after the war to keep their bees alive but sadly most was used to make jam. So the bees that produced honey and left little in their brood chambers for themselves died of starvation during the winter and so over a 20 year period many beekeepers were left with black bees that were useless genetically. The only hives that survived were the ones that gave no honey to their owners. Thereafter the beekeepers were given sugar that contained very persistent green dye and then there was green honey for tea!

After the disruption of the war many honey farms were started by people with little experience using mainly Italian bees and these were susceptible to nosema and acarine. They also shivered all winter and defecated in the hives. A mess that had to be cleaned up by house bees that became further infected. Great effort was made to sell Italian queens right across the nation and many beekeepers in the north, which was a black bee stronghold, bought Italian queens (the grass is greener!). The consequent outcrosses were very aggressive and just about finished village beekeeping. So not only was there a disease problem that got into the black bees there was also aggression and a nutrition problem as white clover disappeared from the pastures to be replaced by Nitram (34.5% N). Selective weed killers were also used that killed clover. Indigenous bees were no longer indigenous although there would be some black bees left in remote areas. The only way to get rid of nosema etc was to hope that all susceptible colonies died in the winter. This worked well. After that there were huge losses in the winter of 1963, very often 100%. Packages were bought in from America in cargo planes. These bees were even more susceptible to disease. They were distributed around Northumberland, I remember, by the college advisor. There were French black bees brought into Scotland in considerable numbers to replace losses and produce heather honey. They were no doubt *Apis Mellifera Mellifera* (black bees the same as our bees), but they could be very defensive during manipulation.

There were vast losses in a period extending from the summer of 1985 to spring 1988 owing to one bad winter and three bad summers. Ours lived through that but only just.

More recently there have been ever increasing numbers of imports from many different countries in order to make up for losses due to varroa and viruses and an ever-increasing demand for bees.

Throughout all these difficult times there were beekeepers who could make black bees look very good indeed with minimum selection.

There was a partnership between the bees and the beekeeper. The bees belonged to an individual. As an aside, over selection will cause black bees to become susceptible to EFB.

Nutrition is the most important factor in beekeeping and there are parts of the UK that will hardly support a hive or two in a poor summer and some candy will be needed. Thrift and black bees go together saving the beekeeper a lot of trouble. This is a very important trait not present in imported hybrids. A colony might need 100lb of honey for maintenance per annum before any surplus is gathered. A sobering thought. Black Bees are able to change their strategy to whatever area they find themselves in. A poor area means a small colony. Also their colour, darker the further north one goes, to absorb sunlight and fly in lower temperatures (down to 6°C in the sun). They will also defer to their keeper over the long term provided he or she doesn't make any glaring mistakes i.e. low swarming tendencies and good nature very often without selection. The bees will select themselves. I remember when we were building up numbers 50 years ago, we were using the nine-day inspection and artificial swarms to go forward. The bees became less pleased to see us every time we went (due to over manipulation honeybee colonies need to be left alone). Then we were breeding from colonies that were preparing to swarm which was the exact opposite of what we should have been doing. We stopped all that and their temper has improved to this day. We then introduced some 'wrinkles' to persuade them not to swarm to the extent that now there are too few swarms. Our local pest controller, a man of considerable experience tells us that pre varroa/viruses he went to 20 honeybee swarms per annum and now he is down to 2 on average sometimes none. This is a clear indication of colony health in our region. My father used to say 'if the bees are swarming all is well with the world'. That was after we lost a big swarm! – but he was right.

In writing about black bees, I should mention Willie Smith of Innerleithen who designed the Smith hive and kept black bees for 40 years in Peeblesshire both semi and fully commercially. His black bees came from tiny beginnings to massive colonies occupying 6 BS deep boxes as I was shown. This was a very difficult area and he effectively fed them heather honey by leaving honey in the super combs from the previous autumn. Now these hives were as strong as any in the country then and now, but he never had to bring in any queens from around the world to achieve that. Neither did he seek any recognition for what had been achieved. Willie Smith called his hive the 'Peebles Hive' together with the 'Peebles Honey Press'. Others thought it should be called the 'Smith Hive'. His motto was: 'Be good to the bees and they will be good to you'. Willie Smith was in the Battle of the Somme in 1916.

Current research will show that there are still black bees around that were here 100 years ago. It should also show traces of all the imports that I have mentioned. Black bees need to belong to somebody either a beginner with aptitude or a beekeeper with ability. So it will not be possible to identify them with any degree of certainty. They might be useless or they might be superb. It all depends on their owner and where they are kept.

*Postscript*

This article was written some time ago. Recent research has shown that it is possible to positively identify black bees, but the quality of these bees still depends ultimately on the beekeeper.

# Andrew Scobie obituary

I was very sorry to hear of the death of Andrew Scobbie from Kirkaldy, Fife, aged 87.

Andrew was keeping a few hives of bees in a council house garden in Thornton, Fife just after the war. The neighbours did not complain about the bees and his father looked after them when he was away doing National Service. The Isle of Wight disease was rampant at the time.

Andrew was a coppersmith, making hot water cylinders for industry. Copper sheet was spun on a lathe to make the convex tops and the concave bottoms of the cylinders – very skilled work. He became a commercial beekeeper around 1965, keeping black bees that were imported from France by Steele & Brodie, hive makers from Dundee and settled on 450 hives as a manageable number. He would make the roofs and floors, buying in the brood chambers and frames. He made his own foundation.

He had a queen rearing apiary just outside Kirkaldy, in a very sheltered location. His queen rearing was a great success, just like everything else he did. He kept his colonies on double brood chambers (Smith) and once the queen cells were ready, split them taking the old queen and a box of brood to another location and introduced a queen cell into the brood on the old site. He added another brood box to the old queen and then took the box with the new mated queen and the original, now double chamber colony, to the heather at the Sma' Glen between Crieff and Amulree. There they were reunited into one unit with the young queen put down below the excluder and on to two boxes (contraction). He then had skyscraper colonies at the heather for the sole purpose of obtaining bulk heather honey. In a good season this was a huge success, poorer seasons might require the colonies to be fed. At some stage his black bees became inbred and susceptible, due to him always breeding from the best of the best. We gave him 2 queens of a lowly status and he used these to requeen his colonies with an unrelated strain of black bees. This worked very well and just about cured the problem in one year.

There was a bout of very poor weather between 1985-88. In the winter of 1985/86 his bees wintered 100% being the only commercial beekeeper in Scotland to achieve this. Other less fortunate souls lost all their bees at this time. Andrew then set about restocking all these people, such was his generosity. He would charge them very little and get the money at a later date. He devoted some time to running the Beekeepers Association at Freuchie, together with others. Robert Couston told him he was the best beekeeper in Scotland of his generation.

I am proud to record this testimony to an exceptional beekeeper and craftsman, and a kind and considerate gentleman.

His business continues today in the ownership of Andrew Scobbie Jnr.

*Willie Robson*

# Selby Robson 1905 - 1990

Born into a local family of coal miners in 1905, Selby Robson attended Shoreswood School from where he obtained a scholarship to Berwick Grammar School and, in order to obtain a further scholarship to Durham University, he excelled in the study of Latin and Greek and became Dux of the school. At Durham University he studied Agricultural Botany (organic) working for some time at Cockle Park and Shoreswood as well as teaching infants at Norham School.

Upon graduation he took up teaching agriculture/horticulture in the south where he met and married his wife, Florence Robson.

After the end of the war, they moved north and established a small market garden here at the Chain Bridge with hens, bees and soft fruit. His mother's family lived here before the bridge was built in 1820.

In 1949 he accepted a post with the Edinburgh and East of Scotland College of Agriculture teaching beekeeping. His family always kept bees (60 skeps originally). He was then able to build up an encyclopaedic knowledge of practical beekeeping by studying the successes and failures of some 150 beekeepers that were within his circle of influence. Some of these beekeepers were quite exceptionally competent in a craft of great complexity. This knowledge was essential in building Chain Bridge Honey Farm. I was fortunate to travel around with him as a child and learn beekeeping by association. I started work here in 1963 on a fairly disorganised level whilst making some useful progress. Selby Robson retired from teaching in 1971 and we worked together until he finally retired aged eighty. At that stage he was proud of what had been achieved but always reserved. We would have been 'looking after' more than one thousand colonies at the time.

We did not get on very well (generations!) but still well enough to make steady progress through some difficult times. My Mother insisted that work continue! Giving up or the thought of giving up would not be tolerated.

Selby Robson knew so much about bees that even today I am constantly referring to what he might have said. He would have been very disappointed by the current state of beekeeping as things have become so difficult as a result of disease, pesticides and low morale among the colonies.

# *Willies Memoirs – Early Years*

My father and mother and I came here from Horley in Surrey in 1948, initially staying in a tent in the summer and staying in rented accommodation in Paxton while their house was built in the corner of a field right beside the Union Chain Bridge. My grandmother's family were living near here before the bridge was built in 1820.

Thereafter the ground was given over to strawberries, raspberries, blackcurrants, 100 hens together with 50 hives of bees acquired through time. These were in Smith hives although the family had cottage hives right back to the skep era. Cottage hives were village joiner built with a fixed floor, gabled roof, single-walled brood chamber and double-walled above for the production of sections (comb honey). The cavity was packed with 'auld semets' which were discarded woollen vests and long johns! These hives were very easy to move because they could not come apart.

But I digress. In 1949 my father got a job with the Edinburgh and East of Scotland College of Agriculture, teaching beekeeping. Whilst he knew a lot about bees, having been among them all his life, he would very quickly learn from beekeepers that he visited as part of his job. He would very quickly understand why some were doing better than others. There were a great number of beekeepers in the rural areas at that time, keeping a huge number of colonies. One village of 80 inhabitants had 200 colonies, another small town had 1500 colonies in total. The problems mainly related to swarming, most of the colonies were feral and habitual swarmers. My father taught the beekeepers to clip the queen's wings to stop her leaving with a swarm and then to carry out 9 day inspections. This did not work out very well in practice because these beekeepers mostly worked a 48hr week, often having to go through them in the evening. Thus the bees became extremely defensive. Nosema and acarine (diseases) were endemic at this time and my father told beekeepers never to crush a bee in a colony either between the combs or the supers. To do so regularly would start an outbreak of nosema. Treatments, and there were many, only caused the bees to become more susceptible in the long term. American and European foul brood were unknown at the time other than little lots bought in from the South that were treated and contained. No colonies were ever burned. Fields and fields of white clover meant that every beekeeper got a crop of honey in all but the poorest years.

There were a good number of beekeepers that were able to get their hives to behave in an orderly manner. This was about aptitude and ability. Willie Smith of Innerleithen was easily the most prominent among those that stood out, although even he had some difficult times. My father would learn a lot from him although there were others, Alec Cosser a salmon fisherman from Kelso, was able to control 10 hives with minimal effort. Smith by contrast had 120 colonies.

Thus my father was fortunate to know so many people who thought about bees and very little else!! And I benefitted from this hugely. Although at the time thought nothing of it. I was born to beekeeping!

Meanwhile, my mother and some friends from the village worked away at the market gardening with only moderate success, bad weather and Botrytis resulted in a lot of fruit being wasted. At some stage in the 50's my mother took up teaching again in order to pay for me to spend 10 years at boarding school, up until 1962 perhaps. Being a mummy's boy I didn't take to boarding school very well but stuck it out. There were times when I lost touch with home although my mother wrote every week I rarely wrote back. Holidays were spent walking the countryside with a dog and gaining a thorough understanding of nature, as well as visiting beekeepers. This 'practical' education was to an higher standard than that received in term time, although to be fair I learned a good amount in every subject at public school and retained most of it.

Thereafter came home for good and wondered what to do. My mother said "your father wants to have a proper honey farm" and so began work in a half-hearted way using the original honey house (40' x 15') keeping colonies in Smith hives and moving hives to the clover in Lauderdale and Abbey St Bathans, and Chatton for the heather. Most of the honey was sold to local grocers and Princes St in Edinburgh (Melroses Tea and Coffee). In 1961 my father bought a big Morris van which was useless for beekeeping. We had been using an Austin A40 pick up and as it had been retained my father lay underneath, removed the sump as well as the cylinder head, disconnected the con rods and pushed the pistons up through the engine until he could replace the piston rings, fitted new bearings to the big ends, ground in the valves and bolted it all up together, all in situ and in the open. That engine then ran to 180,000 miles until we bought a new VW pick up in 1966.

During the winter of '63 we had 4ft of snow that lay for 12 weeks. there weren't many hives alive that spring, although the bees would recover fairly quickly. Our farm is situated on the top of a high bank without any shelter from the North and is not suitable for wintering bees. Additionally bees were flying over the river to some parkland during April and May and too many were being lost into the river because of melting snow on the higher ground. A point to watch!

I was in the Young Farmers in Berwick. They were very much involved with stock judging (not me), public speaking and providing entertainment as a competition. Whilst there were dances every weekend during the winter and some fair bouts of drinking, this was a very well run organization. I also joined the rugby club in Duns and again this was well run. The 30 players were all working class lads. The local farmers would not join this club because of class distinctions which was a bit dispiriting. I met in with some entertainers at a beekeeper's social in Caddonfoot and they set me off with some useful comic songs and I was singing solo in the pubs and village halls at this time. I joined the male voice choir in 1968 singing four-part harmony. We had this at school singing anthems and church music. I still go to the choir sometimes although there were times when I found it heavy going because of business commitments and too much line practice. The male voice choir was another great organization embracing lifelong friendships, some of them overseas.

Meanwhile at home attention was given to building hives and shallow frames. My father had a little table saw and timber being obtained as offcuts locally. I remember that in order to make the joints, I packed out the saw blade with cardboard to make it wobble. The blades had to be sharpened with a file, although I never really mastered the saw sharpening. This was the start of a big woodworking expansion to make Smith hives. There was only one single-phase electric supply, so an Armstrong Siddeley diesel engine was purchased with a big old thicknesser that had been used for making coffins on one pulley and a hefty circular saw on the other pulley, all driven by flat belts and in the open. When the belts started to slip we daubed them with black treacle and were covered from head to foot with shavings. The timber came from Thuja trees that I purchased that had been sawn into sizes to suit the construction of the Smith hives. (I recently bought ten tons of Thuja trunks that had blown down during Storm Arwen). This was a simple hive to make with half lap joints throughout. Floors were made from Iroko that was being used in a local boatyard. Both the Iroko and Thuja gave off chronic dust that gave me asthma, from which I haven't recovered.

Making hives with antiquated and underpowered machinery, especially in the open, became too much and we gave the matter our best attention. Another single phase supply was put into the workshop, now with a roof, to give us 480 volt dual phase and 440 volt 3 phase was created using phase changers. This worked to a degree but the electric motors lacked power when asked to work hard. Eventually we were given a big diesel generator by a utility firm and so we could run any machine. Spindle moulders with power feeds were purchased as well as a dust extractor! And by the time everything was up to standard we had made more than a thousand Smith hives, including the frames. Periodically boxes and frames were bought from E H Thorne in the flat.  Now we are making replacements to a very high standard.  Early on we needed hundreds of hive roofs and we made these out of scrap timber using used litho plates from the printers for the coverings.  The roofs were too shallow at 4" deep and now we have made all our roofs 10" deep western red cedar in order to turn the weather off the hive bodies and keep them dry.  The shallow roofs have all been set aside!  A waste of effort but they were cheap at the time!

It was in 1967 that we built a portal framed shed, my friend Murray Kenny in the village cutting up the steel and welding the angles as necessary.  We had no plans or design – just got on and made it to the rule of thumb, as everyone else did at the time, perhaps not everyone.  It was at this time that the business became an obsession, advancing on many fronts.  My ribs were broken at the rugby playing in the front row for Berwick Rugby Club and being 'laid off' had time to think about the future.  I knew that I would get nowhere staying at the rugby, what with injuries and the hangovers.  Forthwith weekends were spent building during the winter and every Sunday in the summer, as well as nights (under floodlight) for a period of 20 years.  A new purpose built honey house and storage buildings were built exceeding 25,000 sq ft and all to a very high standard.  The design became obvious because the building was purpose built in stages for honey production and works well to this day.  Another big honey house was built subsequently which was given over to extracting honey and bottling, whilst cut comb is packed in the original building.  The buildings were all put up using our own labour.  We were very lucky that George Herkess offered to help because he liked work.  He was a plasterer having served his time in the 'thirties' when mistakes and lethargy would result in the sack.  To be entirely truthful it was he that was driving me a lot of the time.  Progress did not stop.  Then there was Tom Edgar, a farm worker on the next farm, who would turn up whenever needed and make cement or whatever .  He helped me move the bees, with a tap on the bedroom window "Are you waken Wull" at 4am.  He was allergic to beestings but never alarmed by it.  His doctor said "The next sting is going to kill him" – he died of old age!

As mentioned my father had bought some bees in the sixties and I was left one hundred colonies in three beekeepers' wills.  We were looking for new sites and increasing numbers, using nine-day inspections and making artificial swarms, although making increase from swarmy bees is not such a good idea, particularly as many of them were feral/section bees.  At some stage we reached 600 colonies having given up on nine-day inspections and continuing to make nuclei.  Initially queens' wings were clipped to stop them swarming but never marked as my father considered this to be an insult to the queen.

We were involved at that time in many other projects relating to a rapidly expanding beekeeping business. My father had put some money into the early years as well as keeping his son but declined to get involved in any further outlay which was a good thing. Thereafter the business relied on "profits retained", backed by the odd bank loan. The bank manager said that he would rely on me to run the business in a sensible manner and there would be no limit placed on my overdraft and no security was required. Grants were avoided because I was sure we could do the 'work' for less money than the sum required to match the grant and the work would be done to a higher standard and a lot of paperwork avoided. At a time when there was 30% inflation, the Inland Revenue gave us our income tax back so we could buy materials that were 30% dearer the next year, another lucky break.

Other mainstream projects included 'cold selling' into shops. On several occasions I could remember coming close to tears having been given the 'dandy hurl' out of a big shop. It was necessary to stand in a queue to speak to the buyer but our honey generally sold so through time they were quite pleased to see me. Others in the queue (reps) got a 'roasting' because their products were slow sellers. At some stage I had about seven shops in and around Princes Street, often delivering over the pavement from a flat pick-up, all the honey being in whisky boxes with divisions or biscuit tins for sections. My father got a telling-off from a policeman for cutting across the path of a tram whilst delivering to Melroses Tea and Coffee in Princes St. Soon after that they scrapped the trams and more recently they put them back! All payments in those days were cash or seven days, no credit was ever demanded.

Quite the worst outcome of the selling job was that days had to be taken off away from the bees and so a lot of swarms were lost through not being there. The losses at the time seemed to be devastating and were, but there was nothing I could do about it. During May and June the bees wouldn't wait and neither would the shops. There were times when I could have given up and got a job but my mother told me to stick to it!

As the buildings went up so machinery had to be purchased to go in them. As a family we had visited Tom Bradford who had 450 hives in Worcestershire, Herefordshire and Gloucestershire. We noticed he had an industrial spin drier to get the honey out of his cappings. The wax cappings came out as dry as sawdust, maybe not quite, and we were still using a heather press which was extremely inefficient. I went to Faslane on the Gare Loch where ships were being dismantled and purchased a big heavy centrifuge that had been used for drying clothes and that would pay for itself the first time we used it, in that it took every last drop of valuable honey out of the wax. Thereafter many trips were made to boat breakers in Inverkeithing and Blyth to get tanks, sinks and workbenches that we still use. We were allowed, my wife Daphne and I, onto the boats to dismantle as necessary with a hammer and chisel!. On another occasion, I helped to dismantle a school kitchen, obtaining some jacketed pans. A stainless spin drier was bought from a boat for £10. It took 30 amps to start it ( a lot) and was extremely dangerous when it went out of balance. The honey had to be exactly the right temperature. I lay over the top of it more than once until it calmed down, otherwise leave the room and come back later when it might have smashed itself to pieces. It was used to get the heather honey out of the pieces of comb that were not suitable for cut comb. No honey was ever wasted after that. A perfect piece of machinery.

When we began to get oilseed rape honey in the early seventies there was a problem with it crystalizing in the comb, making it impossible to extract in a conventional manner. Most of our shallow frames contained starter strips and no wire, so we could leave the honey on the hive so the bees did not starve and deal with it in the winter. I managed to obtain a machine from a local factory which had been used to mash boiled potatoes. This we used to make the solid rape combs into a slurry which was warmed in a hot cupboard until nearly all the crystals had melted. Thereafter the honey could be dealt with normally. This was a poor system, the honey had been heated for too long, much too long, and lacked flavour as a result, and became darker in colour – not good.

I was giving a lecture in Norway to beekeepers through an interpreter and was talking about the difficulties we had with rape crystals (they knew about this), when a fellow in the audience said he had made a centrifuge that was able to differentiate between rape crystals and beeswax flakes and he would make one for me for £1500. He was a retired bus driver. This machine does exactly as he said and has been in continuous use ever since (25 years). The tank was all stainless but the 'welding' was all silver solder. All beekeeping tinware used solder at one time including extractors and setting tanks. This machine 'worked' because of the difference in specific gravity of beeswax (0.8) and honey (1.4). we were then able to crush the combs of rape honey and warm the mixture to 40°C and pump it through this spinner to remove the wax. The wax was used in our cosmetic business.

Thereafter, after a period of settlement we pump the liquid honey and the crystals through a great big homogenizer. I was aware that the French beekeepers were making "creamed" honey by passing it through a roller mill. I bought a granite roller mill out of a scrapyard in N Yorkshire. It had come from Terry's of York (chocolates). A beekeeper friend fabricated new rolls in ½" stainless steel and we were able to break up crystals in rape honey. Unfortunately, the machine drew in air which spoiled the sample. Additionally, the oil seed rape crystals were harder than stainless steel and were wearing the rollers until they were bowed and useless. I sold this machine back into the confectionery trade.

I was also aware that the French were using homogenizers to make smooth honey. I purchased for £200 one from an ice-cream factory in town, also a huge stainless-jacketted tank that my great friend Ted Thomson used for dipping candles. This machine was a 3cyl Weir pump, as used by boatbuilders (on the Clyde), with stainless wearing parts for milk and cream. The honey is forced through a tiny orifice against a serrated disc which is held down by a coil spring set at 3000 psi. Thus the rape crystals are broken up resulting in a sample ready to bottle after settlement. We can now produce a very acceptable product in 3 passes from rock-solid rape honey with only background heat – a result at last. More recently we are not getting mush rape honey because of new varieties and what we do get will extract giving us some useful drawn frames for the heather (no wire). I bought a tangential extractor in Germany and I am greatly impressed with the quality of the engineering. We had built two automotive reversing extractors years before using thyristor drives (before electronics?) We did this out of enthusiasm and the need to extract heather honey. But then we were trying to produce comb honey and still are. We should never have launched into this project and there were others that didn't work out. Our engineering beekeeper friend Murray Kenny insisted on trying things out. Where he worked (Jus Rol) nothing was "off the shelf", all fabricated in-house and in stainless steel – same story here.

I sold some of the failures to a dealer in Trawden called Keith Birtwhistle, who exported them to Pakistan to be recycled into their food manufacturing industry. Other "failures" I have kept, in the certain knowledge that one day I will need them.

# *Willies Memoirs – Our House*

As I started to put up the buildings for beekeeping, I was aware that a house would have to be built as Daphne and I were married in 1973. We designed the house ourselves and plans cost £25 (there wasn't much detail), which were duly passed by the local authority.

A local farmer, Mr Miller, said that if we went down to the river we would get all the gravel we needed and another farmer allowed us into the dunes to get sand. The general idea at the time was to build the house for £5000, money which was borrowed from the bank.

Initially there were three of us working, George Herkess, a skilled man, Tom Edgar, a labourer and myself, learning as I went. We set all the levels I remember by laying a long garden hose out over the site and propping one end up at the level required and then filling the hose with water which gave us a correct level everywhere at the other end of the hose. Otherwise the house (70' x 30') was set out using triangulation and diagonals.

There were no bricks available for the footings due to strikes, so we obtained quality bricks from a demolition site in Cramlington and removed the lime. Then we plumbed the corners and laid to the line and no mistakes or George would go silent!! We obtained 'clinker' blocks from a power station on the Firth of Forth and 20 tons of artificial stone from a granite quarry at Kemnay in Aberdeenshire. This stone was far harder than any natural stone that could be obtained and all cut to size for £2/sq yd. My father said that no timber framed houses were to be built on land belonging to him – he got that right too!

Slowly but very slowly the house took shape, working at nights and all weekend but not so much in the beekeeping season. The floors were made with insulating blocks laid on their side. We went to Blyth where the blocks were made and there were great stacks of them with the corners knocked off, all available at no cost. I didn't want a cold concrete floor so this is what we did. The roof was massive and we bought trussed rafters (prefabricated) for a quick job to keep the weather out. I would have preferred a joiner built roof. The rendering on the outside was crushed sea shells from the Isle of Barra. I went to Oban to collect the bags of cockle shells. George plastered the inside of the house in double quick time. The ceilings were t and g hardwood gleaned from the boat yard in Eyemouth and all machined up here in the workshop, as were the skirtings and architraves. Daphne, my wife, and I did this work that took for ever and we were covered from head to foot with fine dust, together with cedar dust, which caused me to become asthmatic.

Once I got the ceiling up and the nails punched in, I scraped the surface of the hardwood with pieces of plate glass to make the surface shine, as I was told to do by a local cabinet maker, Jim Turner, who also built us a big table for the kitchen. Unfortunately, the house took us ten years to finish, we went into one room at a time because our son Stephen was born. Previously we had lived in a little hut on site. Whilst building the house, a great deal of attention was given to beekeeping and selling and delivering honey. Building was also taking place in the beekeeper premises with additions every year or so in order to accommodate expansion and all paid for with retained profits from beekeeping. The site was very untidy for more than 20 years with piles of stone, sand, timber and mud in the winter. This was a bit depressing.

The house has turned out to be a considerable success with no faults of any consequence. This would be down to luck and having time-served men to help, especially George. The building inspector, whose name was Bobby Hall, designed the drains and chimney. He had spent the war as a Japanese prisoner of war on the Burma Railway.

Our buildings were put up to a good standard which gives us some degree of satisfaction. Sometimes the work was too much, especially with the uncertainties of beekeeping going on simultaneously. The house cost more than £5000, as time went on we built in some good quality fixtures and fittings.

# *Willies Memoirs – The Bees*

As I explained, my father bought bees, some good and some not so good but none with any disease fortunately. We were left hives that were from feral stock, by John Hall, who wished us well. Whilst I was grateful for his kindness I had a great deal of trouble with them because they weren't commercial colonies. We have seen that feral colonies (amm) want above all else to stay as feral colonies!

After the winter of 1963, there weren't many bees alive in Northumberland and very many packages of American/Italian bees were brought into the UK. Our friend, Hodgson Gray, received some that became degenerate because they could not stand the weather at 56 degrees North. We were left these also and had a difficult time with them. Either you have to kill all the queens in the Spring and buy-in mated queens or given time they will pull themselves together, generally a long time. Robert Couston's hives (Principles of practical beekeeping) came here as a gift and they were very good as I well remember.

We never felt inclined to buy any bees in from abroad. A friend was rearing Buckfast bees nearby and we had a few as a gift but they were very nice bees and the outcross to a black bee also but at every visit they had to be fed during mid summer, whilst black bees kept on storing honey by ceasing to rear brood. This gives the beekeeper and the bees some security in a difficult season. We thought about keeping some Carniolans at one time but news of their swarming activities put us off. New strains of Carniolan bees have been improved in this respect. Very recently a small sample of our bees tested 90-94% Apis mellifera suggesting that left alone black colonies will predominate, that is the ones that will do best in this climate.

We would not consider changing our bees to a different strain now despite the fact that we have had such troubles with varroa and viruses. Eventually they will get through these trying times.

Very recently we have gone into a friendly partnership with Michael Collier who is a highly professional queen rearer. He hopes to turn his little business into a full-time occupation. He uses our breeder queens to rear queen cells for us to make nuclei, as well as a few for himself. Assessment of the breeder queens is made visually by looking at the quality of the hive stock that she breeds. Black colour is important but sometimes the colonies are very dark brown but still of the same pedigree. Resistance to chalk brood is very necessary. These bees have always been highly resistant to Nosema, acarine and EFB, maybe AFB as well. I cannot be sure. I find this new queen rearing development very interesting and gratifying. The seasons here are very short which makes queen rearing difficult, but they intend to rear queens here as well. We are too close to the north-east coast, with perishing cold winds, sometimes still blowing into mid June!

# Willies Memoirs – 1985

1984 was a record season with perhaps 30 tons of honey held over to the following season. 1985 by contrast was desperately difficult. The sun never shone for the whole season and bees at the heather persisted in dying from starvation. I took fondant out to them weekly and spread open packs around among the hives but some colonies had not got the confidence to come out and get it. I only lost about 10 hives but the psychological effect was not good for my wellbeing, wondering where to go next. The bees got nothing that year. Fortunately most of them got through the winter because I fed them fondant which acted as a stimulant to make them breed. Sugar syrup does not do this and it encourages robbing as well. That winter 90% of the colonies in Northumberland died, as well as Scotland. Our bees were featured on national tv having survived in great numbers. I suspect our colonies would be free of nosema as well, which would make a big difference. It was easier for us to get the bees through the winter than it was the preceding summer. 1986 was little better with 10 tons of poor quality honey being produced. Every living thing needs the sun. 1987 and we were back to 1985 weather. All the summer was spent trying to keep bees alive. We were down to 24lbs of honey for the shops that summer. No beekeepers had any honey to sell. Many of them had no bees. Most of the big commercial men to the north of here packed up for a year or two, after three bad years some for good.

1988 came along and we got more than 30 tons of honey. What surprised me was that the bees were all up and running as if nothing had happened. During 1986 and 1987 when there was hardly a sunny day, they had all superseded their queen and presented themselves ready to go. These were colossally resilient colonies, always there or thereabouts with barely and empty hive to be found by mid summer – they are not like that now. They need constant assistance from humans just to survive. Things have come full circle, an epidemic no less.

I've no doubt scientists are right about global warming but we have seen appalling weather over my lifetime. Flooding on the Tweed was far worse 50 years ago than it is now and then 1975-76 particularly '76 resulting in the most persistent drought (18 months). The bees didn't mind so much as they would get water somewhere (dew) but the plants were all dead so no honey, none at all.

# Willies Memoirs – Supermarkets

I was delivering honey into a shop in St Andrews called Geddes, perhaps 35 years ago, where it was quite normal for Mr Geddes to give me £700 out of the till for a delivery of honey. He said to me "I'm finished soon and you are as well" because Wm Low (a supermarket) is opening up next door. This was the beginning of the supermarket invasion where most of the little grocers shops disappeared and town centres became derelict. We soon became short of customers and began stockpiling honey that we could not sell. It could have been about that time that the wholesale price of honey dropped to as low as 50p/lb.

We then had to embark on a branding exercise which continues today and shops had to be found further away. We had a 'round' in Edinburgh which deteriorated to next to nothing, as supermarkets swamped the city. Supermarkets were able to withhold payment to their suppliers for up to 120 days, at a time when the 'trade' adhered to cash or 7 days. This policy was supported by a cross party of MP's who backed their friends in the city rather than their constituents working in agriculture. Supermarkets also destroyed many little businesses that supported the community, because the supermarkets would not and could not pay their bills on time and also made sure that their overseas suppliers in foreign lands got a pittance for their labour. Honey from the Argentine which was of excellent quality fetched 30p/lb delivered into Europe. Some of this honey was sold as "English" by the honey packers!

All this came as a big setback for ourselves but after a great amount of effort being put into marketing, we are trading with more than 400 privately run retail outlets, having lost a few over Covid. Getting through Covid wasn't easy with sales going down by several hundred thousand pounds. Government grants helped, as well as direct marketing through the internet. Where we go from here I cannot say with a 20% increase in costs year-on-year. Certainly honey doesn't want to be much dearer. Uniting a nucleus into every colony before the heather would help to produce bigger crops. Once the cost of keeping a colony is met any honey produced in excess of that cost is free. Most farm animals have to be fed, sometimes a great amount, but black bees do not often have to be fed.

# Willies Memoirs – Cosmetics

Many, many years ago my mother, on her travels with my father to beekeeper meetings in the south, bought hand cream from a lady whose name escapes me, which was of outstanding quality – my mother was never without it, being made with honey and beeswax.

We produce large quantities of heather honey beeswax and my mother suggested we make cosmetics with it. We were originally making a heavy ointment and were faced with the problem of getting honey to mix with the beeswax and almond oil and stay there so that the balm was stable. This we achieved by using a high speed silverson mixer together with controlled heating and cooling. All the machines/tanks were rescued from scrapyards including the filler which was electro-mechanical (out of the ark!). Thereafter emulsions were made using the same basic ingredients and this little business now continues apace after 30 years of production. These cosmetics are very good value for what they are and beekeeper made, which is an added bonus and the public buy them again and again and they are an important source of income.

Quality beeswax is now much more valuable than honey. Imported beeswax is often adulterated and not worth anything, although wax from Africa might be of an acceptable quality. It is not possible to make cosmetics unless the wax is of the highest quality. Beeswax that is used in industry is called BP and everything that might be useful is removed i.e. propolis and aroma. We have learnt a great deal over the years and are confident that these products are up to standard. Heavy polishes are also made here, incorporating carnauba wax.

# *Willie's Memoirs – The Showroom*

At one time visitors came about when we were working in the honey house and work stopped for amiable discussions. Therefore, I thought it might be a good idea to build a proper visitor centre for many good reasons. Firstly, in memory of my father who died in 1990 and to improve beekeeping education, and to provide cashflow which might cancel out our overdraft. We were stockpiling honey from one season and gradually releasing it to the shops the following season and then many shops were reluctant to pay. We were always owed fifty thousand pounds and additionally we were required to pay income tax on money that we hadn't received, as well as make a living.

When we put the big building up we built a second floor for storage which didn't work out too well. Everything needs to be on one level. Thus we converted 200 sq metres into 'education' with a big 12 frame observation hive, together with a vast amount of beekeeping information. Ann Middleditch did the research very thoroughly. A lady called Doreen Irving was looking for a job and she was a trained calligrapher. Her husband was a picture framer, so between them they did the lot. It was interesting how she was able to vary the size of the writing so that everything fitted. Her work is so accurate that not many people would notice that it was hand-written.

Lyndon Purvis, from the village, took care of all the joinery work. Cupboard doors were made in yew wood from some trees that I had bought previously and a massive 2 tier table was made by local cabinet makers from an oak tree that I bought locally. Finally, A D Johnson, a teacher who had taken early retirement was employed to paint pictures on all the spare flat surfaces. One wall has 113 pictures of places that might be seen in a journey up the river Tweed. This showroom is an exercise in traditional skills, to a very high standard and was paid for over an extended period (without grants) and now we are able to sell some of our honey at retail price and are better placed to support the shops. More recently the shops are keener to pay their bills as they have come to realize that we are all in this together.

# Willie's Memoirs – Tractors

We were taking bees to Lauderdale every year to get clover honey, onto a stock farm called Huntington, belonging to the Runcimans. They liked the bees coming there every year and a family friendship built up. I clearly remember whilst working there on the bees, looking up into the sky and seeing an Avro Vulcan bomber passing above. It seemed that these planes were always there. My father might have known what they were about but said nothing. This was the cold war and these planes were on 'red' alert, capable of delivering a nuclear warhead at a moment's notice.

However, the point was that the Runcimans owned a little Caterpillar crawler tractor that had come over from America during the war, having run the gauntlet of the U-boats. This was a vehicle of such quality that we had to buy it. This was the first of many tractors that we rebuilt to 'as new' condition, as an 'exercise'. We had respect for the engineers that built them and for the people that used them.

A beekeeper came to visit us from South Australia, called Ernie Nietzche and when he went home he bought 2 Lanz bulldogs that were parked up near the Flinders Mountains and sent them over in a container. They have both been rebuilt together with many other vehicles, too many to mention. A Routemaster bus has been here for 12 years and is in use for charitable trips. A 1964 Bristol 'Lodekka' has been here for many years and is used as a small café. More recently there has been a lull in restoration work due to beekeeping becoming difficult and less honey being produced. However, I feel that now is the time we should get going again, because the public love to see all these old vehicles, especially when they are up and running. We have a 1906 Railway Carriage to finish along with many others. We are thinking about tackling a 1918 ploughing engine that was condemned by the boiler inspector. Confidence is slowly returning Autumn 2022.

# Willies Memoirs - Conclusions

It was always accepted that 'weather' would be the controlling factor over beekeeping in the North but when the bees aren't healthy beekeeping can become doubly difficult. We have had to buy honey in from fellow beekeepers in the North-East. Our marketing ability has overtaken our ability to produce enough honey to meet the demand. Fortunately, 2022 has been a good year with a great amount of comb honey being produced. We have always concentrated on production of comb honey unlike other beekeepers. The extra revenue has enabled us to pay wages and invest in the business.

We are supposed to be keeping 1600-1800 colonies and did before the varroa epidemic. There have been many empty hives in every apiary in years gone by although not so many this year. This situation has been caused by me hoping that, as in the past, the bees would get over their troubles by themselves. It has taken longer than expected although this year could be a turning point. Twenty years is quite a while for us humans but not for a honey bee that has evolved over many millions of years. Reproduction has broken down within the colonies to a considerable degree in that the queens don't live nearly as long as they used to and the bee's instincts are also compromised. There are times when they don't know what they are doing. Viruses are likely the cause along with pollutants in their brood boxes. Pollutants are everywhere in fact.

At present we are improving the 'amenity' around our premises with time spent in establishing gardens ready for the grand opening of the Chain Bridge which has been totally rebuilt after 200 years and by which time the family will have been 'working' here for 75 years.

Beekeeping is a remarkable industry in that it costs the environment nothing except some fuel to run the vehicles for the occasional visit and yet it gives back so much in terms of pollination and enables us and many others to make a living in the countryside without financial support. Additionally, Chain Bridge Honey Farm still keeps indigenous black bees commercially. These bees are considered to be a 'rare breed' throughout Europe.

Beekeeping and honey production is labour intensive particularly with all the honey and 'beeswax' going to the shops. Hives were built and buildings put up in quieter times. Of particular mention was one Willie Kirkup who was a rabbit catcher to trade as well as a beekeeper. When my father and I went away in the mornings we told Willie what we needed for the next day, and it was always ready at night when we came home. We came to rely on him so much in the early days. Steven Purvis who came here 25 years ago can lay his hand on any job and has become an outstanding beekeeper.

My wife Daphne has helped enormously through a period of 50 difficult and demanding years, as well as bringing up 3 children. Our two eldest children have decided that beekeeping is not for them but our youngest daughter, Frances, contributes more than anybody to the business, as well as dealing with the public. We have some outstanding employees. This at a time, when in my lifetime, the rural economy has collapsed.

This business, like every business and particularly livestock businesses, has been a rollercoaster of ups and downs, mainly downs in recent times like the country itself. I've been here for 60 years with no thought of giving up, not yet anyway.

*Willie Robson - His Words*